Reliability of Power Electronics Devices and Converter Systems

Reliability of Power Electronics Devices and Converter Systems

Guest Editors

Jun Wang
Jun Zhang
Kun Tan
Jingwei Zhang
Hengyu Yu

MDPI

Basel • Beijing • Wuhan • Barcelona • Belgrade • Novi Sad • Cluj • Manchester

Guest Editors

Jun Wang
College of Electrical and
Information Engineering
Hunan University
Changsha
China

Jun Zhang
School of Electrical
Engineering
Hohai University
Nanjing
China

Kun Tan
School of Electrical
Engineering and Automation
Anhui University
Hefei
China

Jingwei Zhang
College of Mechanical and
Electrical Engineering
Hohai University
Changzhou
China

Hengyu Yu
Electrical and Computer
Engineering
The Ohio State University
Columbus, OH
USA

Editorial Office
MDPI AG
Grosspeteranlage 5
4052 Basel, Switzerland

This is a reprint of the Special Issue, published open access by the journal *Energies* (ISSN 1996-1073), freely accessible at: https://www.mdpi.com/journal/energies/special_issues/4135VVTF0W.

For citation purposes, cite each article independently as indicated on the article page online and as indicated below:

Lastname, A.A.; Lastname, B.B. Article Title. *Journal Name* **Year**, *Volume Number*, Page Range.

ISBN 978-3-7258-5385-4 (Hbk)
ISBN 978-3-7258-5386-1 (PDF)
https://doi.org/10.3390/books978-3-7258-5386-1

Contents

About the Editors

Jun Wang

Jun Wang is a professor in the College of Electrical and Information Engineering at Hunan University. His research interests include power electronics and power semiconductor devices. He is an IEEE senior member, an Associate Editor of the *IEEE Journal of Emerging and Selected Topics in Power Electronics*, and Associate Editor of *IET Power Electronics*. He has published more than 100 IEEE journal and conference papers and has more than 40 Chinese patents. He is the recipient of the Second Prize for Technological Invention from the Ministry of Education and First Prize of Innovation Award from the Invention Association.

Jun Zhang

Jun Zhang is an Associate Professor in the Department of Electrical Engineering at Hohai University, where he also serves as a Master's Supervisor and was selected for the 9th China Association for Science and Technology (CAST) Youth Talent Support Program. His primary research focuses on the condition monitoring and thermal management control of new grid-connected energy equipment. Over the past five years, he has led five major research projects, including the National Natural Science Foundation of China Youth Project, the Jiangsu Provincial Natural Science Foundation Youth Project, a State Grid Corporation headquarters science and technology project (as topic lead), and a sub-task of the National Key Research and Development Program of China. He has authored or co-authored over 30 high-quality papers as first author or corresponding author, with more than 10 indexed in SCI and over 20 in EI. Additionally, he is the second author of an English monograph published by Springer Nature and holds three authorized national invention patents as the first inventor. His contributions to research have been recognized with a First Prize Special Award for Science and Technology Progress from the China Power Supply Society as a key contributor, and his doctoral dissertation was honored as an Outstanding Doctoral Dissertation of Chongqing.

Kun Tan

Kun Tan is an Assistant Professor in the School of Electrical Engineering and Automation at Anhui University. His research interests include advanced driving, condition monitoring, packaging, and characterization of power semiconductor devices. He has authored or co-authored more than 40 peer-reviewed publications including more than 10 SCI-indexed journal articles, and serves as a reviewer for leading IEEE and IET journals. He also obtained more than 10 authorized invention patents in China.

Jingwei Zhang

Jingwei Zhang is an Associate Professor in the College of Mechanical and Electrical Engineering at Hohai University. His research interests include modeling of PV module and PV system, failure mechanisms of PV module, fault diagnosis, and intelligent maintenance of PV systems and PV fed microgrids. He has authored or co-authored more than 60 peer-reviewed publications including more than 20 SCI-indexed journal articles, and serves as reviewer for IEEE and Elsevier energy-related leading journals. He also obtained more than 10 authorized invention patents in China.

Hengyu Yu

Hengyu Yu is a Postdoctoral Fellow in the Department of Electrical and Computer Engineering at The Ohio State University, where he is also a member of the OSU ECE Wide Bandgap (WBG) research group. His research focuses on silicon carbide power devices, the reliability and ruggedness of silicon carbide power devices, and their applications in power electronics. He has authored or co-authored more than 40 peer-reviewed publications and serves as a reviewer for leading journals, including *IEEE Transactions on Power Electronics, IEEE Transactions on Electron Devices, IET Power Electronics, Electronics*, and *Micromachines*.

energies

MDPI

Article

Interference Characteristics of Electromagnetic Transient Overvoltage on Secondary Equipment of UHV Fixed Series Capacitors

Baojiang Tian [1], Pei Guo [2,*], Xingwei Du [1], Xiaoyu Liao [1], Chao Xiao [2], Yiran Dong [3] and Jingang Wang [3]

1 State Grid Henan Electric Power Company, Zhengzhou 450052, China; 45464123tbj@sina.cn (B.T.); dxw1575@126.com (X.D.); lxy03711@126.com (X.L.)
2 Electric Power Research Institute of State Grid Henan Electric Power Company, Zhengzhou 450052, China; xc_037@sina.com
3 State Key Laboratory of Power Transmission Equipment Technology, School of Electrical Engineering, Chongqing University, Chongqing 400044, China; 202211021168t@stu.cqu.edu.cn (Y.D.); jingang_023@163.com (J.W.)
* Correspondence: gp_0371@163.com

Abstract: This manuscript addresses the issue of electromagnetic radiation interference experienced by secondary equipment in ultra-high voltage (UHV) fixed series capacitors (FSCs) under electromagnetic transient overvoltage conditions, which cannot be easily determined. To tackle this, a simulation and analysis method for the electromagnetic interference characteristics of secondary equipment is proposed. First, a primary system simulation model of UHV FSC is established, including modeling the platform's multi-conductor system. The electromagnetic transient overvoltage signals between the low-voltage busbar and the high-potential platform are then simulated and analyzed under two conditions: spark gap triggering and disconnector operation. Next, a finite element model of secondary equipment is created to simulate and analyze the electric field distribution of different materials in the area of the measuring box. The shielding effectiveness of the measuring box is calculated using the overvoltage signals at the measuring box location as excitation. This method allows for the simulation of the electric field distribution in the measuring box area for different materials and calculates the shielding efficiency of the measuring box. It effectively simulates the complex electromagnetic environment of secondary equipment, assesses the electromagnetic shielding efficiency of the measuring box, and provides a theoretical basis for analyzing and improving the anti-interference characteristics of the measuring box.

Keywords: FSC; secondary equipment; shielding efficiency; finite element modeling; electromagnetic simulation

1. Introduction

Ultra-high voltage (UHV) fixed series capacitors (FSCs) in power systems can effectively reduce the inductive impedance of long AC transmission lines, significantly enhancing their transmission capacity and improving the grid stability limit. These devices play a crucial role in protecting the power grid during cross-area interconnections and facilitating long-distance clean energy transmission [1–3]. The secondary system of UHV FSCs is essential for monitoring and safeguarding the safe operation of the primary equipment. The layout of the primary equipment on the UHV FSC platform is relatively compact, with secondary equipment such as the platform measurement box and gap triggering control box positioned close to the primary equipment or connecting leads [4,5]. When the spark gap of the UHV FSC is activated and the bypass tandem isolation switch is operated, currents of up to tens of kA can flow through the equipment or connecting leads. This results in a more severe electromagnetic environment for the secondary equipment on the high-potential platform [6,7]. To ensure the proper functioning of the secondary equipment, it is crucial to analyze its electromagnetic interference characteristics under conditions of electromagnetic transient overvoltage.

Most research on fast transient overvoltage in FSC systems simplifies the series-compensated platform to a single node within the system and assumes that the high-potential platform is at an equal potential to the low-voltage busbar when the spark gap breaks down or during switching operations. For example, the authors of [8] simulated and analyzed lightning overvoltage signals by developing an equivalent simplified model of a 1000 kV series compensation device and proposed a lightning intrusion wave protection scheme. Similarly, the authors of [9] utilized an equivalent circuit to simulate and analyze the overvoltage signal generated by the switching operation of the UHV FSC. The authors of [10] introduced a network equivalence scheme to simulate and analyze line splitting, closing overvoltage, and single-phase transient short-circuit experimental overvoltage signals. These studies equivalently modeled the FSC but overlooked the multi-conductor system formed by the series compensation platform and busbar during electromagnetic transients. Since the potentials of the busbar and high-potential platform vary, different overvoltage signals can be generated at different locations. Therefore, it is crucial to perform a multi-conductor modeling simulation of the FSC platform to study the electromagnetic transient overvoltage process comprehensively.

For the electromagnetic interference affecting secondary equipment on the FSC platform, the authors of [11] examine the electromagnetic disturbances in secondary equipment circuits caused by disconnecting switch operations through on-site measurements and characterize the harassment signals. The authors of [12] conduct high-current experiments on the secondary equipment measuring box using a real experimental platform of the FSC to simulate and evaluate the electromagnetic shielding effectiveness of the measuring box under strong electromagnetic fields at industrial frequencies. The authors of [13] develop a secondary system model to simulate and analyze electromagnetic interference caused by primary equipment actions on secondary equipment. These studies primarily focus on the electromagnetic interference effects at industrial frequencies, often considering the conduction coupling of electromagnetic disturbances to the secondary system. However, they do not assess the electromagnetic shielding effectiveness of the measuring box at frequencies beyond the industrial range.

This manuscript conducts multi-conductor modeling of the primary system of the series compensation platform, simulating and analyzing the overvoltage signal between the low-voltage busbar and the high-potential platform. Based on this analysis, the overvoltage signal is used as excitation to simulate the electromagnetic field distribution at the measuring box of the secondary equipment using finite element methods. This approach enables a comprehensive study of the interference characteristics affecting the secondary equipment.

Section 1 provides an introduction to the current status of overvoltage research on high-voltage series compensation devices; in Section 2, the primary system of the series compensation platform is modeled as a multi-conductor, and the overvoltage signals between the low-voltage busbar and the high-potential platform are simulated and analyzed; in Section 3, on the basis of this analysis, the overvoltage signals are taken as the excitation, and the finite element method is used to simulate the electromagnetic field distributions of the measuring box of the secondary equipment; and in Section 4, a summary is made of the whole paper to draw a conclusion.

2. Modeling and Simulation of FSC Primary System

2.1. Modeling of FSC Primary System

The equipment connected to the main primary circuit of the UHV FSC platform includes series capacitor banks, spark gaps, and metal-oxide voltage limiters (MOVs). The UHV FSC platform and primary system structure connection are shown in Figure 1.

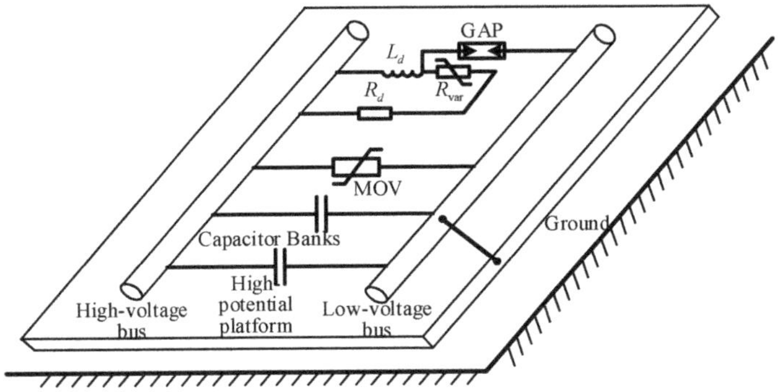

Figure 1. Primary system structure of FSC.

The primary system electrical connection diagram is shown in Figure 2.

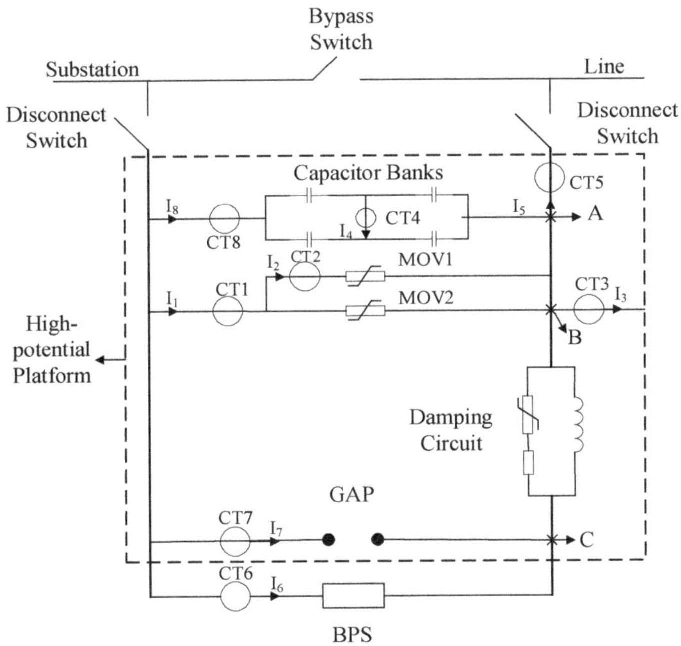

Figure 2. Primary system electrical connection of FSC.

The primary system of the UHV FSC includes leads connecting various components, and under transient conditions, the influence of these leads on the secondary equipment becomes significant. Therefore, it is necessary to model a multi-conductor system that comprises the leads, high-potential platforms, busbars, and primary equipment. The UHV cascade platform can be modeled as a rectangular conductor with dimensions of length L, width W, and height H. To analyze the electromagnetic transient overvoltage at various points on the low-voltage bus and platform, it is necessary to segment the multi-conductor system due to the presence of transverse currents [14]. Figure 3 illustrates the schematic diagram of the platform multi-conductor segmentation and the equivalent circuit of the high-potential platform.

As shown in Figure 3a, the rectangular conductor platform is divided into N_w segments along its width, with each segment having a width of Δw. Consequently, each segmented conductor becomes a rectangular conductor with dimensions $L * \Delta w * H$. In Figure 3b, the resistance of the i-th segmented conductor is denoted by R_i, its self-inductance by L_i, and the mutual inductance between the segmented conductors by M_{ij}. The resistance and inductance parameters for the segmented conductors are derived using the method outlined in study [15]. The capacitance of each segmented conductor to the ground is computed using the finite element method [16].

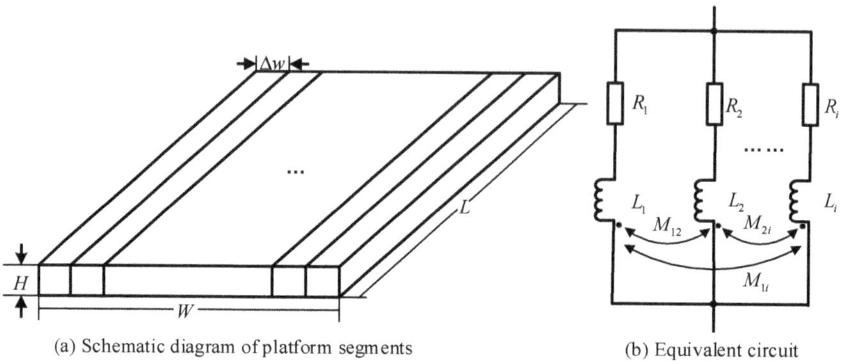

Figure 3. Platform multi-conductor modeling.

On the UHV FSC platform, the low-voltage (LV) buses are initially connected equipotentially to the platform [17]. However, during transient processes, this equipotential condition no longer holds, leading to the generation of fast transient overvoltage at various locations. Consequently, the LV bus must be modeled as a segmented conductor. Figure 3 shows the segmented structure and impedance of the bus.

As depicted in Figure 4a, the high-voltage bus is segmented into H_n sections, while the low-voltage bus is segmented into M_n sections. In Figure 4b, R_{Hi}, L_{Hi}, and M_{Hij} denote the resistance, self-inductance, and mutual inductance of the conductors in the i-th segment of the high-voltage busbar, respectively. Similarly, R_{Li}, L_{Li}, and M_{Lij} represent the resistance, self-inductance, and mutual inductance of the conductors in the i-th segment of the low-voltage busbar. The mutual inductance between the i-th segments of the high-voltage and low-voltage busbars is represented by M_{Bi}. In the electromagnetic transient simulation, vector fitting is used to determine the frequency-dependent characteristics of the test rate resistance, allowing for the creation of a wide-frequency equivalent model of the busbar resistance [18]. The inter-bus mutual capacitance and bus-to-ground capacitance also need to be considered in simulation modeling.

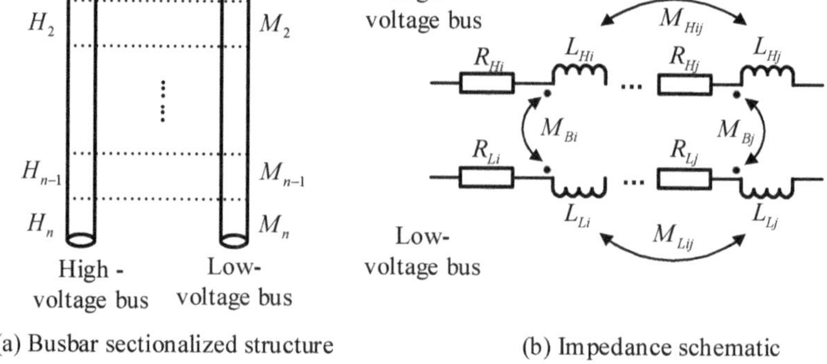

Figure 4. Bus multi-conductor modeling.

The primary equipment of the UHV FSC platform mainly includes metal-oxide varistors (MOVs), spark gaps, series disconnect switches, series capacitor banks, and damping circuits. The series capacitor bank and damping circuit can be represented by their respective equivalent circuit elements. The simplified model of the MOV is derived exponentially from its voltage–current characteristics, as shown in (1).

$$R = \frac{3}{6AT + AT^3u^2} \tag{1}$$

Here, R is the resistance of the MOV; A and T are the MOV volt–ampere characteristic constants; and u is the terminal voltage of the MOV. After the above derivation, the MOV can be equated by a voltage-controlled current source. The equivalent circuits of the spark gap and series disconnect switch in the series compensation platform's primary equipment are shown in Figure 5.

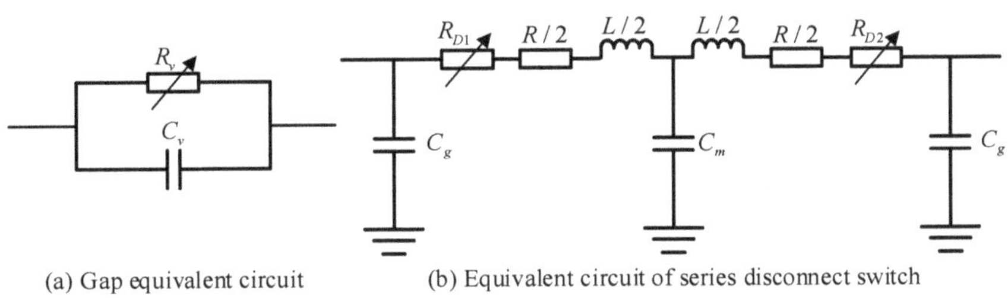

Figure 5. Primary equipment modeling.

As shown in Figure 5a, the triggering process of the spark gap, before it fully conducts, is quite complex and can be represented by a structure where a capacitor is in parallel with a time-varying resistor. In the figure, C_v represents the voltage-equalizing capacitor of the spark gap, and R_v denotes the time-varying resistor, which can be defined by (2).

$$R_v(t) = r_v + R_a e^{-t/\tau} \tag{2}$$

Here, r_v is the stable on-resistance of the spark gap; R_a is the insulation resistance of the spark gap; and τ is the discharge time constant.

Figure 5b shows the equivalent circuit of the UHV FSC platform series-connected isolation switch. R_{D1} and R_{D2} are the fracture switching resistances, which can be expressed by the time-varying resistance in (2); L and R are the inductance and capacitance of the center conductive rod; and C_g and C_m are the ground capacitance of the insulator and the center conductive rod, respectively. The values of some variables of the primary equipment of the FSC platform are shown in Table 1.

Table 1. Primary equipment partial variable values.

Variable	Value	Variable	Value
r_v	0.5–20 Ω	R	0.75 mΩ
Ra	10^{12} Ω	C_g	80 pF
τ	1–20 ns	C_m	160 pF
L	8.5 μH		

2.2. Simulation of Electromagnetic Transient Overvoltage in Primary System

The electromagnetic transient overvoltage simulation model for the primary system is developed based on the actual parameters of a 1000 kV high-voltage series compensation platform. The platform is 27 m long and 12 m wide, and the height of the platform from the ground is 11.87 m. The total capacitance of the capacitor bank is 164.6 μF, with a nominal operating voltage of 98.4 kV. The minimum spark gap triggering voltage is 1.8 p.u., and the MOV overvoltage protection level is 2.3 p.u. When modeling the actual high-potential platform, the platform is divided into nine segments along its width. The locations of the overvoltage measurement points between the low-voltage busbar and the platform are shown in Figure 6.

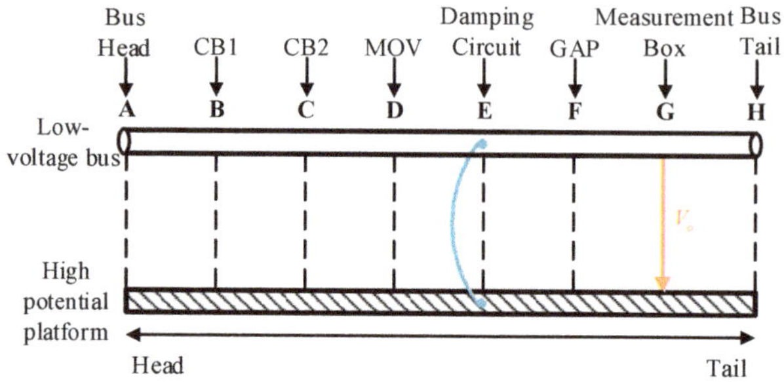

Figure 6. Location of overvoltage measurement points.

The LV busbar is divided into seven sections to simulate the overvoltage at various locations along the platform. Point E represents the connection between the LV busbar and the high-potential platform, while point G is the geometric centerline of the measurement box for the secondary equipment on the platform. The overvoltage signal at point G is denoted as V_0. The electromagnetic transient overvoltage simulation of the primary system primarily focuses on the overvoltage signals between the LV busbar and the high-potential platform under the two most severe conditions: the breakdown of the spark gaps and the closure of the series disconnect switches.

2.2.1. Spark Gap Trigger Overvoltage Simulation

When the spark gap to trigger voltage U_c = 320 kV, stable on-resistance r_v = 0.5 Ω, and discharge time constant τ = 1 ns by the conduction, G point and high-potential platform electromagnetic transient overvoltage waveforms are created between the platform as shown in Figure 7a. The peak overvoltage distribution between the LV bus and the high-potential platform from the head to the tail of the platform is shown in Figure 7b.

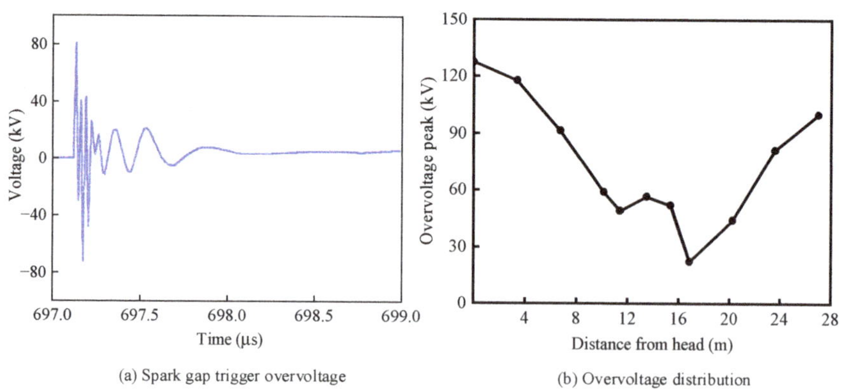

Figure 7. Spark gap triggered transient overvoltage simulation.

As shown in Figure 7a, when the bypass gap is triggered, an electromagnetic transient overvoltage signal of up to 80 kV is generated between the busbar and the platform, which were initially at equipotential. Time–frequency analysis reveals that the dominant frequency of the overvoltage signal can reach up to 107 MHz. Figure 7b indicates that the lowest overvoltage occurs between the low-voltage busbar and the high-potential platform at a distance of 17 m from the first end of the busbar. This lower overvoltage is due to the connection between the low-voltage busbar and the high-potential platform. From point E towards both ends of the busbar, the overvoltage level gradually increases, with the peak overvoltage at the secondary equipment measurement box at point G reaching as high as 81 kV.

2.2.2. Disconnect Switch Operation Overvoltage Simulation

When the series isolation switch is closed, the electromagnetic transient overvoltage waveform between point G and the high-potential platform is shown in Figure 8a. The peak overvoltage distribution is shown in Figure 8b.

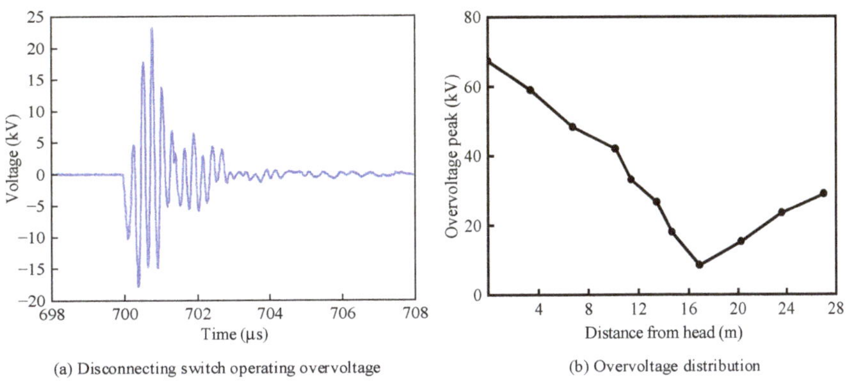

Figure 8. Isolation switch closure electromagnetic transient overvoltage simulation.

As shown in Figure 8a, the amplitude of the electromagnetic transient overvoltage caused by the isolation switching operation is relatively lower than that caused by the spark gap triggering, although its duration is longer. Time–frequency analysis reveals that the main frequency of the overvoltage signal is primarily concentrated around 6 MHz. Figure 8b shows that the overvoltage level generated between the low-voltage busbar and the high-potential platform by the disconnecting and closing operations is significantly smaller than that generated by the spark gap triggering. From the ends of the busbar to the point where it connects to the platform, the overvoltage level exhibits a linear decline, with the peak overvoltage at the secondary equipment measurement box at point G reaching 23 kV.

3. Research on Interference with Secondary Equipment Under Electromagnetic Transient Overvoltage

3.1. Finite Element Simulation Modeling of Secondary Equipment

The finite element model of the secondary equipment on the series compensation platform includes the low-voltage busbar, the measurement box, the PCB board inside the box, and the steel structure housing the box. The simulation model is shown in Figure 9.

In Figure 9, the open seam of the measurement box is indicated by the blue line. The low-voltage busbar is simulated as an antenna composed of two good conductors, with the busbar made of aluminum alloy and having an outer diameter of 70.02 mm. It is positioned parallel to the steel structure and the measurement box. The measurement box has dimensions of 1540 mm in height, 800 mm in width, and 300 mm in thickness, with a material thickness of 2.5 mm. Slits are located on the right side and along the upper and lower edges of the box, with each slit having a depth of 2.5 mm. The box material has a relative permeability of 500 and a relative permittivity of 0.1. Inside the box, the PCB board is simplified as a dielectric layer with the same thickness as the PCB, having a thickness of 4 mm and a relative dielectric constant of 2.4. The steel structure is made of 10-gauge steel. Finite element simulation of the secondary equipment is completed using COMSOL

6.2, and the RF microwave transient module is used to simulate the electromagnetic wave transient field on the surface of the measuring box of the secondary equipment.

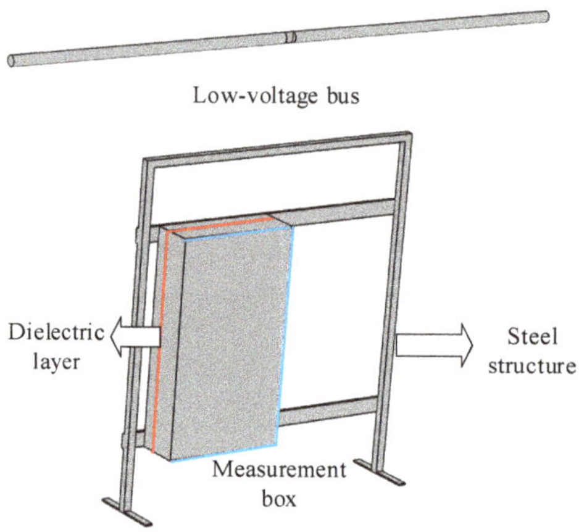

Figure 9. Secondary equipment finite element model.

3.2. Distribution of Electric Fields in the Area of Secondary Equipment Shielding Devices

Through the simulation analysis in Section 2, the electromagnetic transient overvoltage signals between the low-voltage busbar and the high-potential platform were obtained from the secondary equipment measurement box under two different scenarios: spark gap triggering and disconnecting switch operation. These two distinct overvoltage signals were used as excitation sources for the low-voltage busbar to simulate the complex electromagnetic environment around the secondary equipment measurement box and its electric field distribution. The electric field distribution on the surface of the measurement box at a specific moment during the electromagnetic transient process is illustrated in Figure 10.

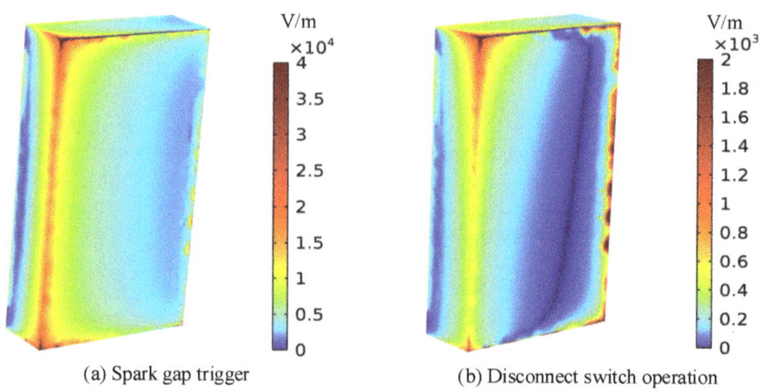

Figure 10. Electric field distribution on the surface of the box.

As shown in Figure 10, under spark gap triggering conditions, the electric field intensity on the surface of the measurement box exhibits a broader distribution, reaching up to 32 kV/m. The electric field gradient is more pronounced, with the highest intensity concentrated around the open slit of the box. In contrast, under the operating conditions of the isolation switch, the charge distribution is more uniform, reducing electromagnetic radiation leakage. The electric field intensity on the surface of the measurement box is significantly lower, with a maximum of only 1.98 kV/m. The distribution range of the electric field is narrower, but the region with the highest intensity still occurs around the open slit, which is a weak point in the box's shielding effectiveness and should be given special attention during design optimization. The electric field distribution in the area of the PCB board inside the box is shown in Figure 11.

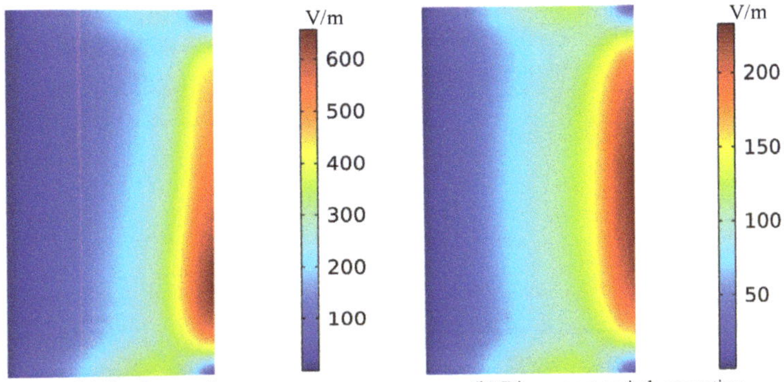

(a) Spark gap trigger (b) Disconnect switch operation

Figure 11. PCB board area electric field distribution.

As shown in Figure 11, the highest electric field intensity in the PCB board area for both cases is concentrated on the right side, primarily due to the open slit in the measurement box, which results in increased electromagnetic radiation leakage. Therefore, when installing the PCB board, it should be positioned away from the side with the open slit in the box. The electric field intensity in the PCB board area is lower compared to that on the surface of the measurement box, indicating that the measurement box provides effective shielding.

3.3. Shielding Efficiency Analysis

The PCB board inside the measurement box contains a large number of low-voltage circuits, which are highly susceptible to electromagnetic interference. This interference can lead to inaccurate measurements by the secondary equipment on the cascade platform, thereby compromising the safe operation of the cascade platform's equipment. Therefore, it is crucial to conduct a more detailed analysis of the shielding effectiveness of the secondary equipment measurement box.

The shielding effectiveness of the measuring box can be calculated by comparing the electric field strength at a certain point on the PCB with or without the measuring box, and the formula for calculating the electromagnetic shielding effectiveness is as follows:

$$SE = 20\lg(\frac{E_0}{E_c}) \quad (3)$$

Here, E_0 is the electric field strength at a certain point of the PCB board when no measurement box is added, and E_c is the electric field strength at that point after the measurement box is added.

Using the electromagnetic transient overvoltage signal generated by the spark gap triggering as an excitation signal, a comparative analysis was conducted on the shielding effectiveness of the measurement box made of copper versus the material specified in Section 3.1. The shielding effectiveness curves of the measurement box at different frequencies are shown in Figure 12.

As shown in Figure 12, when copper is selected as the material for the measurement box, it demonstrates better shielding efficiency in the 0–100 MHz band. There is no significant difference in shielding efficiency between the two materials in the 100–150 MHz band, likely due to the high conductivity of copper. Additionally, it can be observed that the shielding effectiveness of both materials is higher than 100 MHz compared to within the 100 MHz range, with the shielding effectiveness at 100 MHz reaching 66.7 dB.

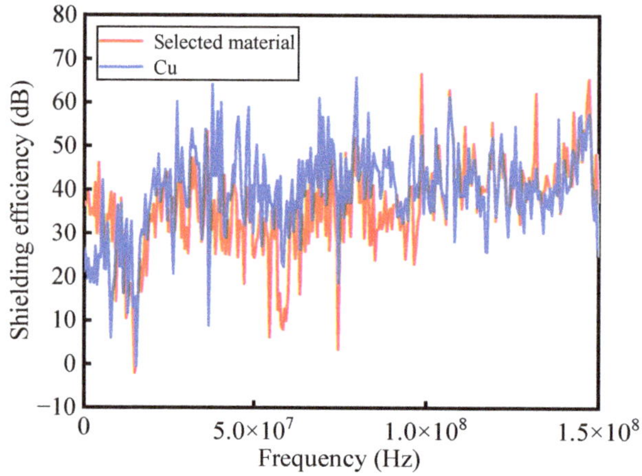

Figure 12. Shielding effectiveness of measuring boxes made of different materials.

To evaluate the impact of dielectric plate material and thickness on the shielding effectiveness of the chassis, a control group is established as detailed in Table 2. This control group is used to analyze the shielding performance across various dielectric plate parameters.

Table 2. Dielectric plate parameter selection.

Group	Dielectric Plate Thickness (mm)	Relative Permittivity
1	4	2.4
2	4	4.7
3	4	9.2
4	6	2.4
5	8	2.4

The shielding effectiveness of the measurement box for different dielectric plate parameters is shown in Figure 13.

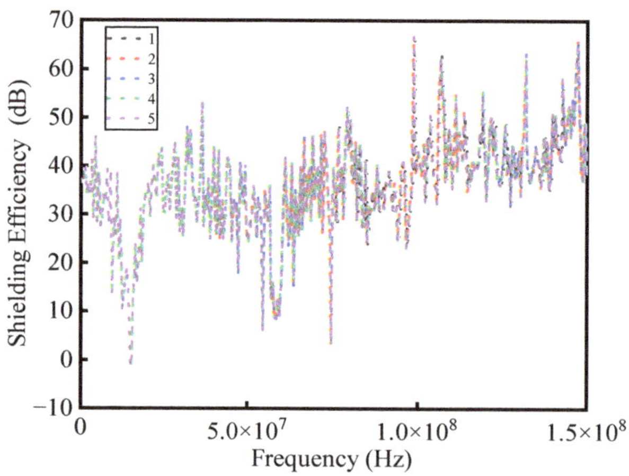

Figure 13. Measurement box shielding effectiveness with different dielectric plate parameters.

As shown in Figure 13, when the thickness of the dielectric plate remains constant, altering the relative dielectric constant of the dielectric plate has minimal impact on the shielding effectiveness of the measurement box. Similarly, when the relative dielectric constant is kept constant and the thickness of the dielectric plate is varied, there is no significant change in the shielding effectiveness of the measurement box.

4. Conclusions

This manuscript presents a simulation and analysis method for evaluating the electromagnetic interference characteristics of secondary equipment in extra-high-voltage string-patching devices. First, a primary system model of the crosstalk platform is developed based on actual engineering parameters. Electromagnetic transient overvoltage signals are simulated and analyzed under two conditions: spark gap triggering and isolation switch operation. Next, a finite element model of the secondary equipment is constructed, where the overvoltage signals between the low-voltage busbar and the high-potential platform are used as excitation sources. The electric field distribution within the measurement box and its shielding efficiency are also examined. The principal conclusions of this study are as follows:

(1) The simulation results indicate that the overvoltage generated by the spark gap triggering and the isolating switch operation is minimized at the connection point between the low-voltage busbar and the high-potential platform. Thus, it is advisable to place the secondary equipment measurement box at this location.
(2) The electric field intensity on the PCB board is highest near the open seam of the measurement box, so the PCB should be placed on the side farthest from the seam. Additionally, the measurement box exhibits better shielding effectiveness above 100 MHz, and its shielding performance improves in the 0–100 MHz range as the conductivity of the box material increases.
(3) The dielectric material and thickness have no significant effect on the shielding efficiency, so materials with higher conductivity are recommended for the measurement box to protect the secondary equipment on the series compensation platform from electromagnetic radiation interference in complex electromagnetic environments.

Author Contributions: Conceptualization, B.T. and P.G.; methodology, X.D.; software, X.L.; validation, C.X.; formal analysis, Y.D.; investigation, B.T.; resources, P.G.; data curation, Y.D.; writing—original draft preparation, Y.D.; writing—review and editing, P.G.; visualization, B.T.; supervision, J.W.; project administration, P.G.; funding acquisition, P.G. All authors have read and agreed to the published version of the manuscript.

Funding: This research was funded by the Science and Technology Project of State Grid Corporation of China: 5100-202424025A-1-1-ZN, research on key technology for reliability improvement of ultra/extra-high voltage series complementary secondary system.

Data Availability Statement: The data used in the analysis presented in the paper will be made available, subject to the approval of the data owner.

Conflicts of Interest: Authors Baojiang Tian, Xingwei Du, Xiaoyu Liao, were employed by the company State Grid Henan Electric Power Company. Authors Pei Guo, Xiaoyu Liao were employed by the company Electric Power Research Institute of State Grid Henan Electric Power Company. The authors declare that they have no known competing financial interest or personal relationships that could have appeared to influence the work reported in this paper.

References

1. Han, X.; Sun, X.; Chen, H.; Qiu, N.; Lyu, D.; Wang, N.; Wang, X.; Zhang, J. The Overview of Development of UHV AC Transmission Technology in China. *Proc. CSEE* **2020**, *40*, 4371–4386+4719.
2. Liu, T.; Li, F.; Lu, P.; Qiao, Z.; Yang, Z. Fault Analysis and Improvement of Optical Isolated Terminals in UHV Series Compensation Control and Protection System. *Electrotech. Electr.* **2023**, 30–33.
3. Chen, Y.; Zheng, Y.; Jiang, C. Study on Electromagnetic Field Distribution of 500kV Series Compensation Platform and Temperature Rise of Fence. *Electr. Eng.* **2022**, *178–179*, 183.

4. Ma, Q.; Cui, X.; Hu, R.; Zhang, W.; Liu, H.; Zhan, X.; Yu, J.; Liu, H. Electromagnetic transient interference measurement system of UHV Fixed Series Capacitors at high potential and EMI characterization. In Proceedings of the International Symposium on Electromagnetic Compatibility, Rome, Italy, 17–21 September 2012.
5. Liu, T.; Guo, G. Analysis of spark gap refusal trigger accident in UHV fixed series compensation and corresponding solutions. *Electr. Power Autom. Equip.* **2021**, *41*, 218–225.
6. Qi, B.; Li, Y.; Ma, S.; Li, B.; Yu, J.; Da, K. In-depth Study on the Contact Fault of Ultra-high Voltage SC Measurement System. *J. Electr. Eng.* **2020**, *15*, 106–113.
7. Ma, Q.; Cui, X.; Hu, R.; Zhang, W.; Liu, H.; Zhan, X.; Yu, J. Experimental Study on Secondary System Grounding of UHV Fixed Series Capacitors via EMI Measurement on an Experimental Platform. *IEEE Trans. Power Deliv.* **2012**, *27*, 2374–2381. [CrossRef]
8. Gao, F.; Chen, W.-J.; Li, G.; Liu, Z.-F. Protection of Series Compensation Station Against Lightning Intruded Wave in1000kV AC Transmission System. *High Volt. Eng.* **2010**, *36*, 2199–2205.
9. Zhou, Y.; Peng, J.; Wang, G.; Zhou, C. Close—Operation overvoltage of UHV AC transmission line with series compensation. *Electr. Power Autom. Equip.* **2008**, *28*, 23–26.
10. You, G.; Si, D.; Chen, S.; Wang, X. 500kV series complementary electromagnetic transient overvoltage study. *Yunnan Electr. Power* **2011**, *39*, 9–11.
11. Li, X.; Zhong, Y.; Li, B.; Huo, H. Measuring and Analysis of Electromagnetic Interference of Switch Operation in 500kV Series Compensation Station. *J. Microw.* **2010**, *26*, 124–127.
12. Zhan, X.; Dai, C.; Liu, H.; Wang, Y.; Yan, H.; Wu, Q.; Cui, H.; Tang, J. Study on High Current Experiments of Measurement Box Located on Platform of Series Capacitor Compensation for UHV Power Transmission Project. *Power Syst. Technol.* **2012**, *36*, 26–31.
13. Ma, Q. Research on Electromagnetic Transient Process in 1000-kV UHV Series Capacitor Banks Due to Switching Operation. Ph.D. Thesis, North China Electric Power University, Beijing, China, 2013.
14. Zhao, P.; Wang, J.; Xia, H.; He, W. A Novel Industrial Magnetically Enhanced Hydrogen Production Electrolyzer and Effect of Magnetic Field Configuration. *Appl. Energy* **2024**, *367*, 123402. [CrossRef]
15. Paul, C.R. *Inductance: Loop and Partial;* IEEE: Piscataway, NJ, USA, 2010; p. 220.
16. Paul, C.R. *Analysis of Multiconductor Transmission Lines;* IEEE: Piscataway, NJ, USA, 2008; p. 780.
17. Zhao, P.; Wang, J.; He, W.; Xia, H.; Cao, X.; Li, Y.; Sun, L. Magnetic Field Pre-Polarization Enhances the Efficiency of Alkaline Water Electrolysis for Hydrogen Production. *Energy Convers. Manag.* **2023**, *283*, 116906. [CrossRef]
18. Zeng, R.X.; Sinsky, J.H. Modified rational function modeling technique for high speed circuits. In Proceedings of the IEEE MTT-S International Microwave Symposium, San Francisco, CA, USA, 11–16 June 2006.

energies

MDPI

Article

Research on Junction Temperature Smooth Control of SiC MOSFET Based on Body Diode Conduction Loss Adjustment

Junke Wu *, Yunpeng Wei, Yuntao Wu, Zhou Wang, Xingyu Li and Xiangnan Wei

School of Mechanical and Electrical Engineering, Guilin University of Electronic Technology, Guilin 541004, China; 20102239@ceic.com (Y.W.); yuntao_wu@mails.guet.edu.cn (Y.W.); wangzhou@mails.guet.edu.cn (Z.W.); xingyu_li@mails.guet.edu.cn (X.L.); weixiangnan@mails.guet.edu.cn (X.W.)
* Correspondence: allenwjk@guet.edu.cn

Abstract: In a converter of actual working condition, the change in the current and voltage of the power device will cause the junction temperature to fluctuate greatly. This device is subjected to high thermal stress due to the change in the junction temperature. Therefore, it is necessary to adopt junction temperature control to reduce or smooth the junction temperature fluctuation, so as to realize the junction temperature control and improve the reliability of the device. At present, the methods for the junction temperature control of power devices have certain limitations and there are few active thermal management methods proposed for SiC device characteristics. In this paper, a method for realizing the smooth control of the junction temperature of a SiC device based on the conduction loss adjustment of the body diode for the SiC device has been proposed, considering that the conduction loss of the body diode is greater than the conduction loss of the SiC MOSFET. The conduction time of SiC MOSFET body diode was adjusted. By adjusting the conduction loss of the SiC MOSFET device, the fluctuation range of the junction temperature of the SiC MOSFET device was controlled, the smooth control of the junction temperature of the SiC device was realized, and the thermal stress of the device was reduced.

Keywords: SiC MOSFET; power semiconductor; junction temperature smoothing control; electric vehicle motor drive inverter; body diode; conduction loss

1. Introduction

Wide band-gap semiconductor materials such as silicon carbide (SiC) and nitride (GaN) have significant advantages in their physical and electrical aspects compared to ordinary silicon (Si) materials. The breakdown electric field strength and band gap of SiC materials are eight times and three times than that of ordinary silicon material [1]. Therefore, SiC material is more suitable for making high-temperature, high-frequency, anti-radiation high-power devices. Power devices made of SiC materials help to improve the efficiency and power density of power electronic systems, and are now widely used in the next generation of high-voltage high-power converters.

SiC MOSFET is the core component of the converter, which plays a decisive role in the safety and reliability of the converter. It is necessary to design a more reliable SiC converter to improve the reliability and prolong the service life of the SiC MOSFET.

The main reason for the failure of power devices is the thermal stress caused by the fluctuation of the junction temperature; the fluctuation of the junction temperature will lead to different degrees of expansion in different layers, resulting in thermal stress impact, which will lead to the fatigue aging of the power devices [2]. The junction temperature fluctuation of the power device is mainly related to the low-frequency junction temperature fluctuation, which is mainly caused by the large-scale random fluctuation of the input and output power of the power converter [3]. The junction temperature of power devices is closely related to their life; according to the device life model studied by LESIT, large temperature fluctuations have a much greater impact on a device's life than small temperature fluctuations [4,5]. Therefore, it is very necessary for power devices to use effective control methods to achieve junction temperature smoothing control to improve their reliability and prolong their service life [6].

At present, there are several methods for the junction temperature smoothing control of power devices. The first method is the control method based on drive regulation, including the junction temperature control method based on gate current regulation [7], the junction temperature control method based on the three-stage gate drive circuit [8], the junction temperature control method based on the gate resistance network [9], and the junction temperature control method based on the switching frequency [10]. The variable switching frequency may affect the output waveform quality of the converter. For this reason, some scholars have proposed the concept of the life improvement benefit index, which takes into account the adjustable range of the switching frequency and the waveform quality.

The second method is the control method based on the changing modulation strategy, including the junction temperature control method based on space vector pulse width modulation [11] and the junction temperature control method based on hybrid modulation [12]. At present, some scholars have proposed that the finite control set model predictive control in modern control theory can be used to realize the junction temperature control of power devices, and this method has great potential [13]. For this reason, some scholars have proposed multi-parameter optimization through the finite set model's predictive control junction temperature [14,15], changing the device's modulation mode, stabilizing the load current, and reducing the junction temperature fluctuation.

The third method is improving the control method of the buffer circuit, including the method of realizing the junction temperature control based on changing the discharge path of the buffer circuit [16]. In the traditional RCD buffer circuit, an auxiliary switch is added to realize the junction temperature control by changing the discharge path of the buffer circuit. When the auxiliary switch is closed, the buffer capacitor discharges to the power device to increase the loss of the power device. When the auxiliary switch is disconnected, the purpose of reducing the loss of the power device is achieved. When the device is turned on, the peak current flowing through the drain source increases. For this reason, some scholars have pointed out the limitations of this method, and proposed a method of changing the buffer mode to achieve junction temperature control [3]. The conversion of the RC buffer circuit and the RCD buffer circuit is realized by connecting the auxiliary switches. By means of loss compensation, the loss of the power device is kept constant to achieve the purpose of smoothing the junction temperature.

It can be seen from the current research that power device junction temperature control research has achieved a certain adjustment effect, but the control method based on drive adjustment has great limitations, and the design scheme needs to be changed according to different power device models. The method of realizing junction temperature control by changing the modulation strategy to stabilize the load current cannot control the junction temperature of a single power device. The control method of the improved buffer circuit has small limitations and simple control. It can be applied to different power devices and is the optimal choice for current junction temperature smoothing control. However, this method requires an additional circuit, which may reduce the efficiency of the converter. At present, there are few active thermal management methods proposed for SiC device characteristics, and this method does not require additional circuits.

This paper presents a SiC MOSFET junction temperature control method based on lossless buffer switch trajectory adjustment. By changing the conduction time of the SiC MOSFET body diode, the conduction loss of a SiC MOSFET device is adjusted, so as to control the junction temperature of the SiC MOSFET device. The feature of this method is that it does not require an external circuit and can only be applied in converters that require dead time.

2. Conduction Characteristics of SiC MOSFET's Body Diode

Due to the contact between the p-well and the source metal, a pin-type diode structure is formed through the p-well, n-region, and n + substrate. The specific structure of the SiC device is shown in Figure 1 and the equivalent circuit diagram of the SiC device is shown in Figure 2. Therefore, for a bridge topology application in a voltage source conversion circuit, there is essentially a freewheeling diode. The characteristics of this diode correspond to the third quadrant of the I-U characteristic curve of the SiC MOSFET. The manufacturing technology of the SiC MOSFET usually leads to a higher carrier lifetime. Therefore, a higher storage charge and higher diode peak reverse current will be generated in the traditional SiC MOSFET. This is an obstacle for many applications of SiC MOSFET devices. The adjustment of the carrier lifetime can be used to reduce the stored charge, but this must be performed through a separate production step.

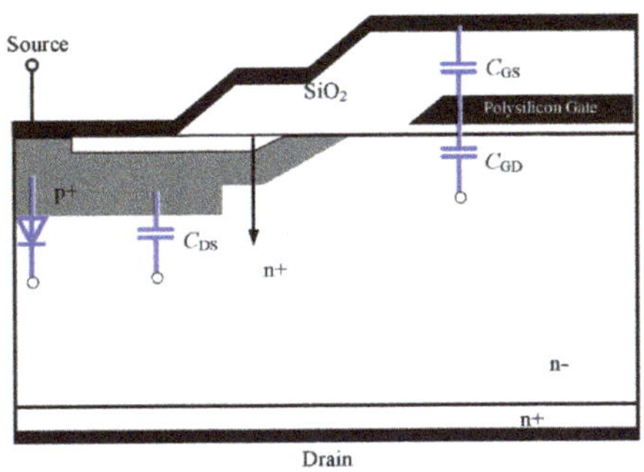

Figure 1. The schematic diagram of a SiC device structure.

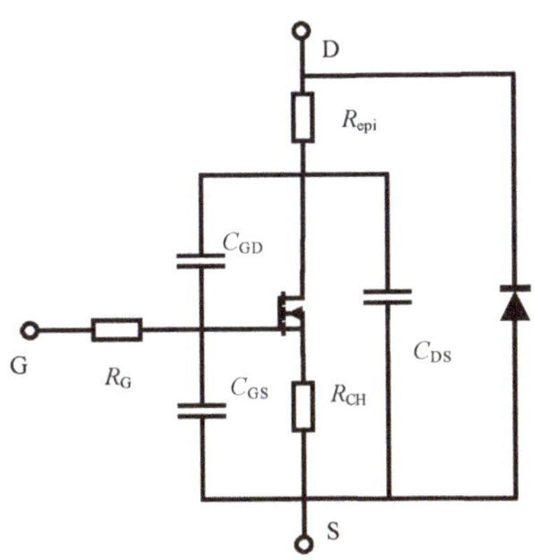

Figure 2. The equivalent circuit diagram of a SiC device.

When the driving voltage at both ends of the gate source is given to −4 V, the body diode of the SiC MOSFET is turned on. At this time, the holes in the P + region and the Pbase region of the SiC MOSFET and the electrons of the substrate will be transported to the N-Drift region due to the applied voltage. At this time, a saddle-like carrier concentration gradient is formed in the N-Drift region, which leads to the weakening of the drift motion, thus generating a diffusion current. The forward current of the bulk diode consists of four parts, namely, the recombination current J_{n-} and J_{n+} generated by the space charge region at the interface of the P + region and the N + region entering the N-Drift region, the electron diffusion current J_{p+} and J_n entering the P + region and the N-region, and the hole diffusion current J_p entering the N-Drift region. The formula is [17]

$$J = J_{n-} + J_{n+} + J_n + J_{p+} + J_p \quad (1)$$

When a large forward voltage is applied to the SiC MOSFET, the concentration of the non-equilibrium minority carriers injected into the drift region is much larger than the intrinsic carrier concentration. The N-type drift region generates the same concentration of non-equilibrium majority carrier to maintain electrical neutrality, so the hole and electron diffusion currents generate positive currents and the body diode obtains a lower on-state voltage drop. The forward voltage drop of the body diode is composed of the voltage drop of the P + contact region, the voltage drop of the drift region, the PN junction voltage drop composed of the P + region, the N + region, the N-Drift region of the body diode, and the voltage drop of the substrate region. The forward voltage drop of the bulk diode can be expressed as follows:

$$U_F = U_p + U_D + U_{p\_n^+} + U_{p\_n^-} + U_s \quad (2)$$

In the formula, U_F is the voltage drop when the body diode is turned on. U_{p-n+} and $U_{p\text{-}n\text{-}}$ are the voltage drop of the PN junction formed by the P + region, the N + region, and the N-Drift region of the body diode. U_s is the voltage drop of the substrate region, U_p is the voltage drop of the P + contact region, and U_D is the voltage drop of the drift region.

The resistance of the drift region is related to the conductance modulation effect and the voltage drop of the drift region is affected by the resistance of the drift region. Therefore, the voltage drop of the drift region is mainly analyzed because the conductivity modulation level is related to the carrier lifetime.

$$k_W = \frac{W}{2L_a} \quad (3)$$

In the formula, W is the length of the drift region, and L_a is the diffusion length. The pressure drop in the drift region can be expressed as follows according to the size of k_W:

$$U_D(k_W < 1) = \frac{3kTk_W^2}{q} \quad (4)$$

$$U_D(k_W \geq 1) = \frac{3kT}{8q}e^{k_W} \quad (5)$$

In the formula, q is the elementary charge, T is the absolute temperature, and k is the Boltzmann constant. The diffusion length is determined by the carrier lifetime in the neutral region. The diffusion length can be expressed as follows:

$$L_a = \sqrt{\frac{\mu_n \mu_p}{\mu_n + \mu_p}\tau_H} \quad (6)$$

In the formula, τ_H is the injected carrier lifetime, μ_n is the electron mobility in the bulk diode, and μ_p is the hole mobility in the bulk diode.

3. Junction Temperature Control Based on Body Diode Conduction Loss Adjustment

Since the band gap of the SiC material is three times that of Si material, the PN junction turn-on voltage and forward conduction voltage drop of the SiC MOSFET are higher than those of the Si-based devices, resulting in the conduction loss of the body diode of SiC MOSFET device being greater than the conduction loss during operation. Therefore, the working time of the body diode of the SiC MOSFET affects the loss of the whole chip.

3.1. The Principle of Conduction Loss Adjustment

Figure 3 is the body diode volt–ampere characteristic curve of the Si MOSFET and the SiC MOSFET. In the Figure, the loss of the SiC MOSFET body diode is much larger than that of the Si MOSFET device, and the conduction loss of the SiC device body diode is greater than that of the device conduction. Therefore, the loss of the whole chip can be controlled by controlling the working time of the SiC MOSFET body diode, so as to achieve the purpose of the junction temperature control.

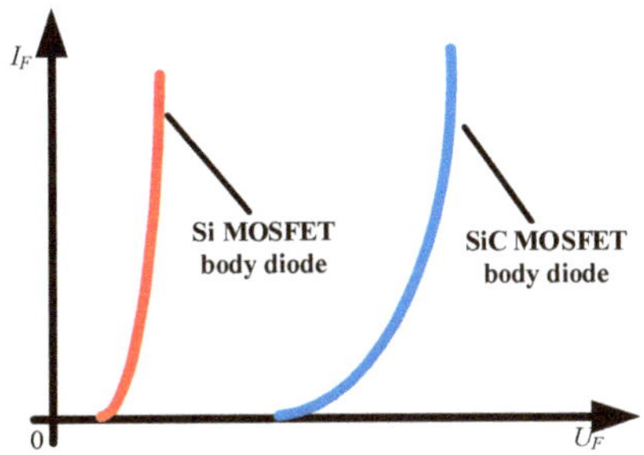

Figure 3. The volt–ampere characteristic curves of body diode of Si and SiC MOSFET.

When the inverter is working, the upper and lower switches of the same bridge arm are forbidden to be connected at the same time in order to avoid a short circuit, and there must be a certain dead time for the upper and lower tube control. During the dead time, when the upper and lower tubes are turned off at the same time, the current flows through the reverse diode of the SiC MOSFET. Due to the large turn-on voltage of the SiC MOSFET body diode, the dead time of the upper and lower tubes can be adjusted; that is, the turn-on time of the SiC MOSFET body diode can be adjusted in a variable direction, so as to realize the adjustment of the conduction loss of the SiC MOSFET, so as to realize the junction temperature equalization control.

The whole commutation process is described by taking the left bridge arm as an example. In Figure 4a, the upper tube is turned on and the lower tube is turned off, and the current flows through the upper tube. In Figure 4b, the upper and lower tubes are turned off at the same time. Since the current direction cannot change abruptly at this time, the current flows through the body diode of the lower tube, resulting in a freewheeling phenomenon. In Figure 4c, the middle and lower tubes are turned on and the upper tube is turned off, and the current flows through the lower tube; the freewheeling process of the upper tube diode in Figure 4d is similar to that in Figure 4b. Based on this principle and these characteristics, a research method based on SiC MOSFET body diode conduction loss adjustment is proposed.

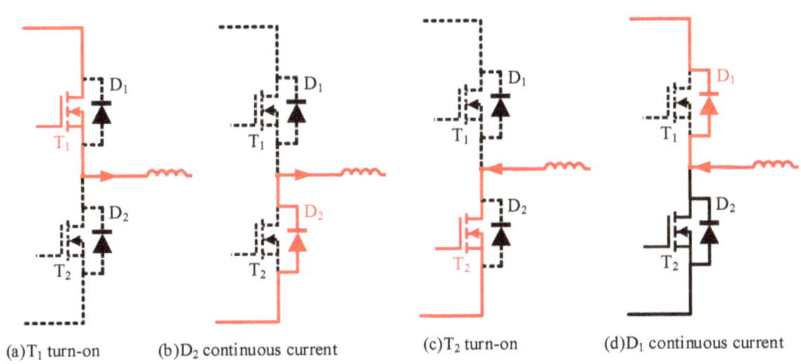

Figure 4. The SiC body diode freewheeling process in the inverter.

3.2. The Influence of Body Diode Conduction Time on the Operation of the Electric Drive Inverter

In the adjustment of the body diode conduction time in the inverter is the adjustment of the dead time of the upper and lower tubes of the same bridge arm. The insertion of the dead time ensures that the switching tubes on the same bridge arm will not be turned on at the same time, which improves the safety of the circuit. However, the dead time will also have some adverse effects on the converter.

When the dead time is set too large, the distortion rate of the PWM waveform will increase, and the output voltage and current of the inverter will produce serious distortion. If the dead time is set too large, it will lead to the distortion of the magnetic chain in the motor control system, which will affect the normal operation of the electric drive inverter. When the dead time is set short, the probability of the switch tube on the same bridge arm of the inverter will increase, which will produce a certain degree of additional power loss, resulting in the heat loss control of the switching device, reducing the physical life of the components in the electric drive inverter, reducing the reliability of the electric drive inverter, and reducing the overall stability and work efficiency of the electric drive inverter control system. Secondly, the dead time also has an effect on the output current of the inverter. When the SVPWM modulation strategy is adopted, if the motor stator current is close to zero in the dead time, the zero current clamping phenomenon will occur due to the addition of the inverter dead time, resulting in the distortion of the current waveform and the fluctuation of the torque. The nonlinear output characteristics of the inverter will affect the stability of the induction motor U/F control system. In addition, the dead time also affects the phase angle of the output voltage. The insertion of dead time leads to an increase in many of the harmonic components in the output waveform of the inverter. The increase in the harmonic components will inevitably lead to the increase in the motor loss, and a large ripple component will be generated in the output torque of the motor, which will cause the oscillation of the motor and the instability of the system.

In the formula, n is the harmonic number, U_o is the harmonic voltage, T_D is the dead time of the inverter, U_d is the pulse voltage, k is the pulse number of half a cycle, and f is the fundamental frequency. According to Formulas (3)–(7), when U_d, k, and f are constant, the harmonic content of the output voltage is related to the dead time. The greater the dead time, the greater the harmonic content of the output voltage; the THD (total harmonic distortion) of the output current is also larger [18].

$$\sum_{n=3,5,\ldots}^{\infty} U_o^2 = \frac{1}{\sqrt{2}}\sqrt{\sum_{n=3,5,\ldots}^{\infty}\left(\frac{4kT_D U_d f}{n\pi}\right)^2} \tag{7}$$

Therefore, it is necessary to compensate the dead zone reasonably; as shown in Table 1, setting too long of a dead time will lead to an increase in the THD of the output current and reduce the conversion efficiency of the inverter. In recent years, a variety of effective compensation schemes have been proposed. For example, the hardware circuit is used to detect the pulse width of the switch tube in real time, and is then compared to the command pulse width to obtain the deviation voltage for compensation. However, these measures may not be able to compensate for the impact of nonlinear loads.

Table 1. The influence of the change in dead time on a THD.

Current I_L	Dead Time t_D	THD
16 A	500 ns	0.15%
16 A	1000 ns	0.30%
16 A	2000 ns	0.60%
16 A	5 μs	1.52%

In general, the influence of dead time on the inverter is mainly reflected in waveform distortion, current distortion, stability reduction, and so on. Therefore, it is necessary to set the dead time reasonably and take appropriate compensation measures to reduce its adverse effects.

4. Experimental Verification of Body Diode Conduction Adjustment

4.1. The Influence of Body Diode Conduction Time on the Operation of Electric Drive Inverter

The flow chart of active thermal management method for SiC MOSFET devices based on body diode conduction loss adjustment is shown in Figure 5.

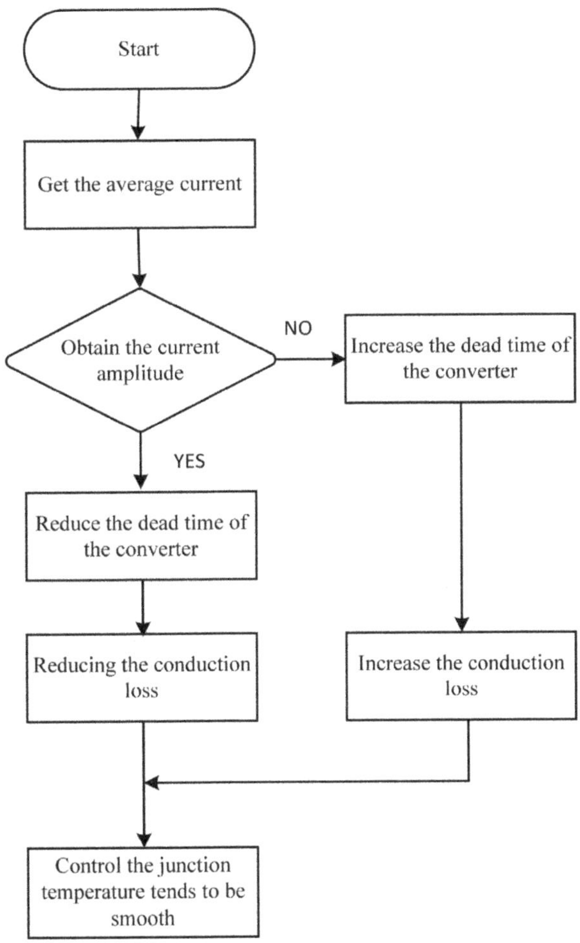

Figure 5. Flowchart of active thermal management of SiC MOSFET.

The experimental verification of body diode conduction loss adjustment is carried out and a simulation verification is carried out on the software Saber. The version number of Saber is 2016. The type number of the SiC MOSFET used in this paper is SCT3030AL, which is produced by Rohm, The Spice model of SiC MOSFET is imported through the device and the inverter simulation platform is built. In the stable operation state, the output current of the inverter load side is 20 A. Table 2 shows the inverter parameters of the simulation design.

Table 2. Inverter simulation parameters.

Symbol	Parameter	Values
U_{dc}	input voltage	200 V
f_m	output frequency	50 Hz
f_w	switching frequency	50 kHz
L_0	load side inductance	1.22 mH
L_{ms}	stray inductance	40 nH

During the simulation process, the load current of 0–0.34 s. 0.68–1 s (*T*1 and *T*3) is maintained at 18 A. The load current of the 0.34–0.68 s (*T*2) cycle is reduced to 16A. At this time, the circuit does not add body diode conduction loss control. In the simulation period of 0–1 s, the inverter dead time is 500 ns. Figure 6 shows the load current and device simulation junction temperature waveform in this state. The upper half of the figure is the load current waveform and the lower half is the device junction temperature waveform of the upper tube of the left bridge arm of the inverter.

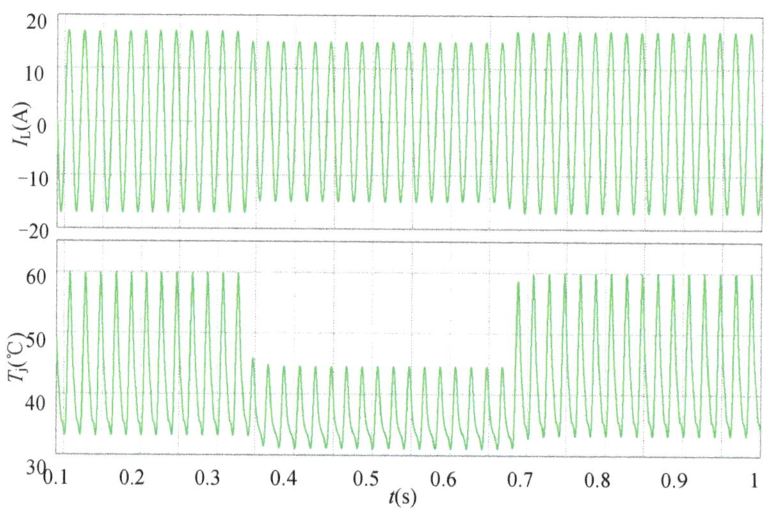

Figure 6. The simulation waveforms of the load current and device junction temperature when the dead time is 500 ns under the inverter commutation.

The body diode conduction loss adjustment is performed for the inverter in this state. In the high current stage (0–0.34 s. 0.68–1 s), the dead time of the converter is reduced to 50 ns to reduce the conduction time of the body diode of the SiC MOSFET device and achieve the purpose of reducing the conduction loss of the SiC MOSFET device. In the small current stage (0.34–0.68 s), the dead time of the converter is increased to 5 μs to increase the conduction time of the body diode of the SiC MOSFET device, so as to increase the conduction loss of the SiC MOSFET device. Table 3 shows the closing parameters of each period before and after the simulation. Under the method of adjusting the conduction loss of the body diode, the change in the junction temperature and the change in the load current are shown in Figure 7. The upper part of the figure is the load current waveform and the lower part is the junction temperature waveform of the device under test.

Table 3. The closing parameters of each period before and after simulation.

Time Interval	Dead Time t_D	Current I_L	THD
Before control T_1 and T_3	500 ns	18 A	0.13%
Before control T_2	500 ns	16 A	0.13%
After control T_1 and T_3	50 ns	18 A	0.01%
After control T_2	5 μs	16 A	1.52%

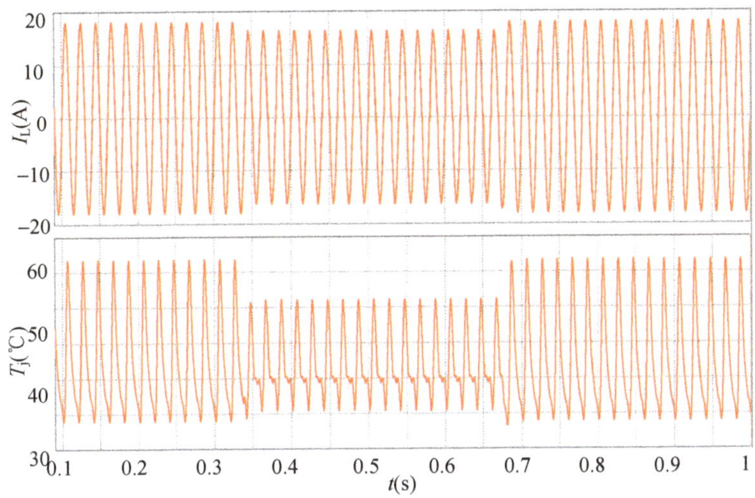

Figure 7. Simulation waveforms of load current and device junction temperature after body diode conduction loss adjustment control.

Due to the lower reduction in the dead time, the junction temperature does not decrease significantly during the small load period. During the large load period, the conduction loss increases due to the increase in the conduction time of the body diode, and the junction temperature of the device increases significantly. However, it can be seen that during the large load period, the load current is distorted. This is exactly what is mentioned in the previous section. The increase in the dead time will lead to the distortion of the load current and the generation of higher harmonics.

Figure 8 shows the waveform comparison before and after the body diode conduction loss control. It can be seen from the figure that after the body diode conduction loss control, when the load current is 18 A, the junction temperature of the device after body diode conduction loss control is slightly lower than that before the body diode conduction loss control in the 0–0.34 s and 0.68–1 s periods. When the load current is 16A, the junction temperature of the device after the body diode conduction loss control is slightly higher than that before the body diode conduction loss control in the period of 0.34–0.68 s. The junction temperature of the device tends to be smooth after control, and the proposed method of the active thermal management of the SiC MOSFET by adjusting the conduction loss of body diode is feasible.

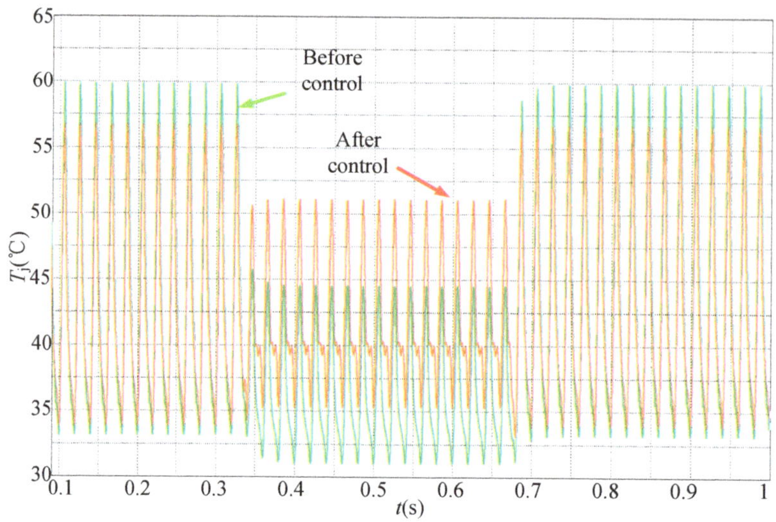

Figure 8. Comparison of device junction temperature simulation waveforms before and after body diode conduction loss adjustment control.

4.2. Experimental Verification Based on Body Diode Conduction Loss Adjustment Method

In order to further study the adjustment method of body diode conduction loss, experiments are carried out on the basis of the simulation platform of Figure 9 to simulate the operating conditions of electric vehicles.

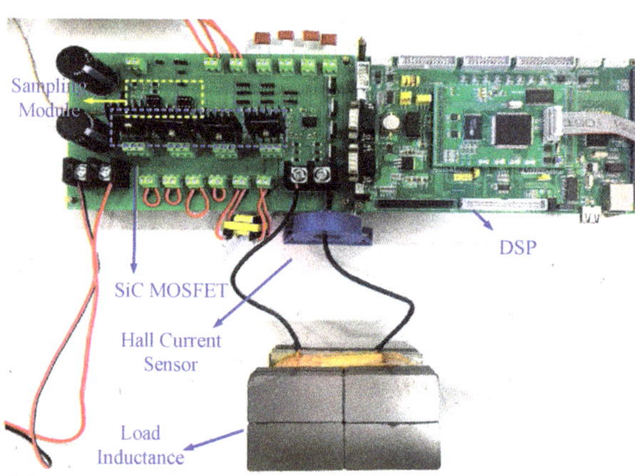

Figure 9. Electric vehicle working condition simulation platform.

The converter control was realized by simulating the operating conditions of the electric vehicles, and the current size was changed every 350 ms. The control current was changed according to the law of 9 A-8 A-9 A and the dead time was 350 ns. Figure 10 shows the load current waveform under the condition of the current change. In this figure, the load current in the large current stage was 9 A, and the load current in the small current stage was 8 A. Under this condition, the junction temperature estimation was carried out, and the junction temperature waveform of the measured device is shown in Figure 11.

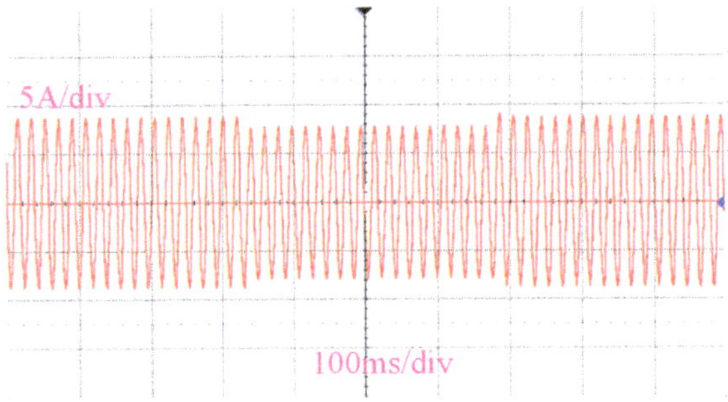

Figure 10. The load current of the experimental platform is 9 A-8 A-9 A.

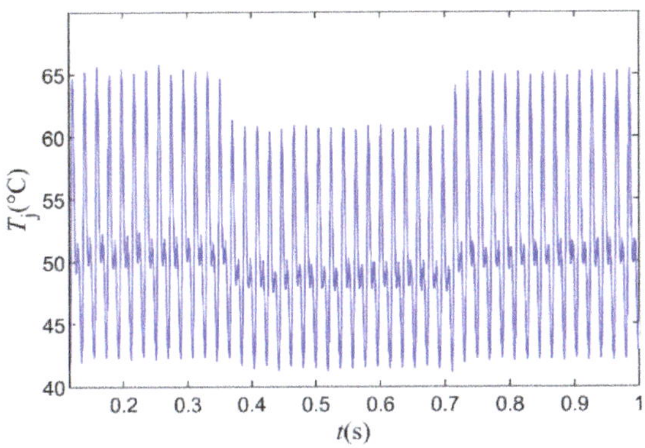

Figure 11. Junction temperature waveform of the device under the test before the conduction loss control of the body diode.

Under the above conditions, the body diode conduction loss control experiment was carried out. In the small current stage, the body diode conduction time of the device under test increased and the dead time was 2 μs. In the large current stage, the body diode conduction time of the device under the test was reduced and the dead time was 100 ns. Table 4 shows the values of the closing parameters in each period before and after the experiment.

Table 4. The closing parameters of each period before and after simulation.

Time Interval	Dead Time t_D	Current I_L
Before control T_1 and T_3	300 ns	9 A
Before control T_2	300 ns	8 A
After control T_1 and T_3	100 ns	9 A
After control T_2	2 μs	8 A

Figure 12 shows the comparison of the waveforms before and after the body diode conduction loss control. The blue waveform in the figure is the junction temperature waveform of the device before the body diode conduction loss control and the pink waveform in the figure is the junction temperature waveform of the device after the body diode conduction loss control. It can be seen from the picture that when the load current is 9 A, the junction temperature of the device is lower than that before the control. When the load current is 8 A, the average junction temperature of the device is similar to that when the load current is 9 A. The junction temperature of the device tends to be smooth after control.

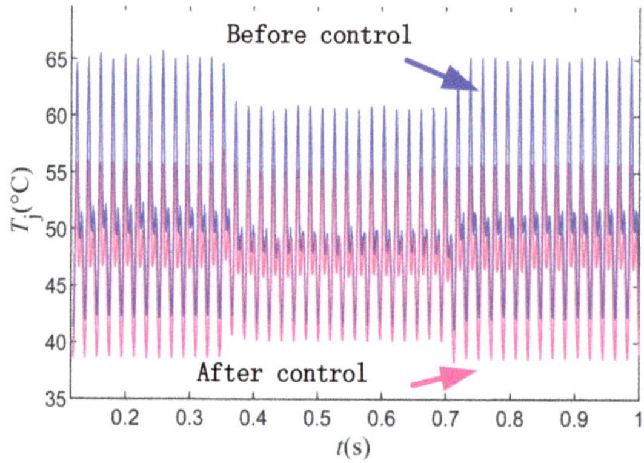

Figure 12. Comparison of the junction temperature waveforms of the device under the test before and after body diode conduction loss control.

After the body diode conduction loss control, the load current was almost unaffected in a certain control range of the dead time, and the load current would not be distorted. Figure 13 shows the load current after the body diode conduction loss control. Figure 13a is the whole current waveform. Figure 13b is the waveform amplified in the high current stage, with an amplitude of 9 A, and Figure 13c is the waveform amplified in the low current stage, with an amplitude of 8 A. It can be seen from Figure 13 that the load current will not be distorted when the dead time does not exceed a certain value.

On the basis of the previous experiment, the dead time of the high current stage is maintained at 100 ns, and the dead time of the low current stage is increased from 2 μs to 3 μs. Table 5 is the value of the shutdown parameters for each period after the experiment.

Table 5. The closing parameters of each period before and after the experiment.

Time Frame	Dead Time t_D	Current I_L
After control (first and third part)	100 ns	9 A
After control (second part)	3 μs	8 A

The load current waveform after the increase in the dead time in the small current stage is shown in Figure 14. Figure 14a is the whole current waveform. Figure 14b is the waveform amplified in the large current stage, and the load current amplitude is 9 A without distortion. Figure 14c is the waveform amplified in the small current stage. The load current amplitude is less than 8 A, and the load current is distorted.

It can be seen from Figure 14 that the dead time is increased to 3 μs in the small current stage, and the load current is distorted from 8 A to below 8 A in the small current stage. Adjusting the dead time can adjust the conduction time of the body diode to change the junction temperature of the device, but too much dead time will lead to distortion of the current which will affect the use of the converter.

4.3. Control Effect Evaluation Based on Body Diode Conduction Loss Adjustment Method

In order to further explore the influence of junction temperature on the life of SiC MOSFET devices, and to evaluate the proposed active thermal management method of SiC MOSFET devices based on body diode conduction loss adjustment, the effect of improving the reliability of the device before and after the control of the on-body diode conduction loss adjustment method was evaluated. The lifetime of the device is closely related to the junction temperature of the device. For the statistics of the junction temperature of the device, the ordinary single-parameter counting method cannot meet the load cycle characteristics. The method for counting the junction temperature of the power device is generally the rain flow counting method.

The rain flow counting method (tower top method) is widely used to extract the load during the cycle. At first, the rain flow counting method was only considered to be used in material mechanics. The rain-flow counting method can take both the mean value and the fluctuation amplitude into account during the operation. This method divides the whole process into many arithmetical thermal stress fluctuations. In this case, the number of cycles corresponding to each level is drawn, and the fatigue cumulative cycle data are calculated by the number of cycles.

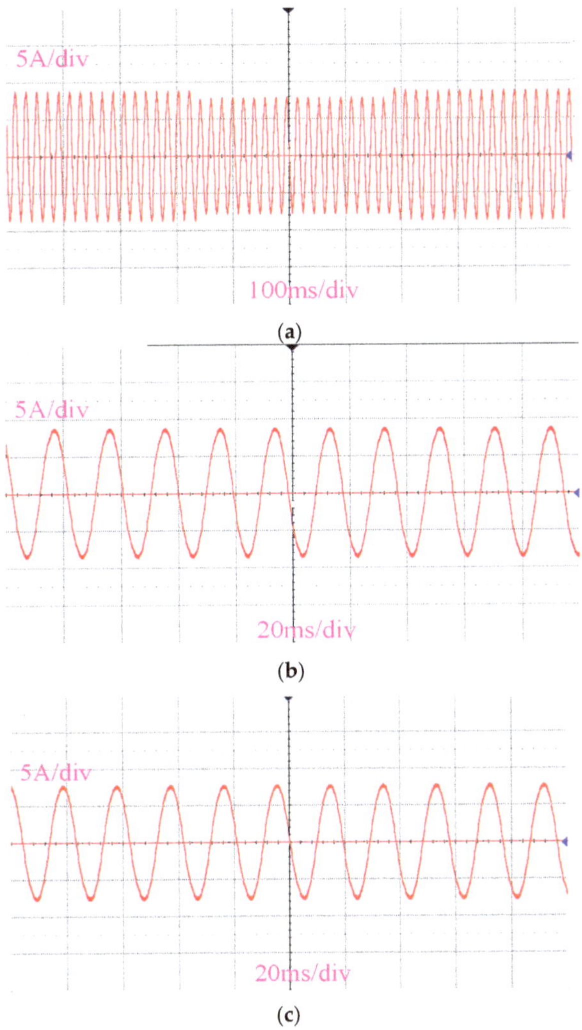

Figure 13. Load current after conduction loss of body diode (dead time of small current is 2 μs). (**a**) The whole current. (**b**) Amplification in high current stage amplification. (**c**) Amplification in the low current stage.

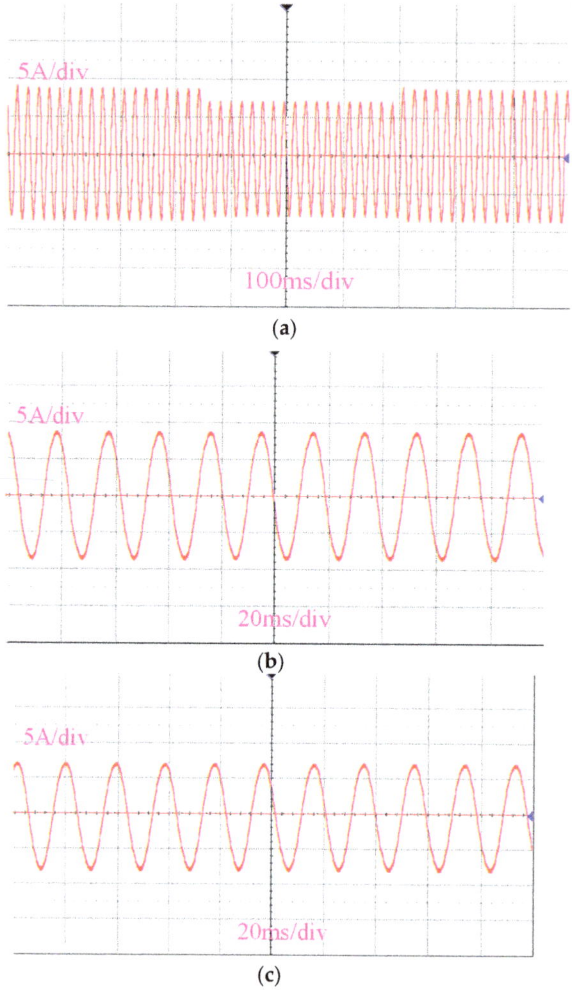

Figure 14. Load current after conduction loss of body diode (dead time of small current is 3 μs). (**a**) Whole current. (**b**) Amplification in large current stage. (**c**) Amplification in low current stage.

Figure 15 shows the heat load histogram of the device under the test before the conduction loss control of the body diode processed by the rain flow counting method. Figure 15 shows the heat load histogram of the device under the test after the conduction loss control of the body diode processed by the rain flow counting method. The main factors affecting the life of a measured power device are the amplitude of the junction temperature change ΔT_j and the average junction temperature $T_{j.m}$. m. The estimation of the number of failed thermal cycles N_f is mainly based on these two parameters. The lifetime of the device is predicted by the junction temperature amplitude, the mean value, and the fluctuation times of each device obtained by the rain flow counting method.

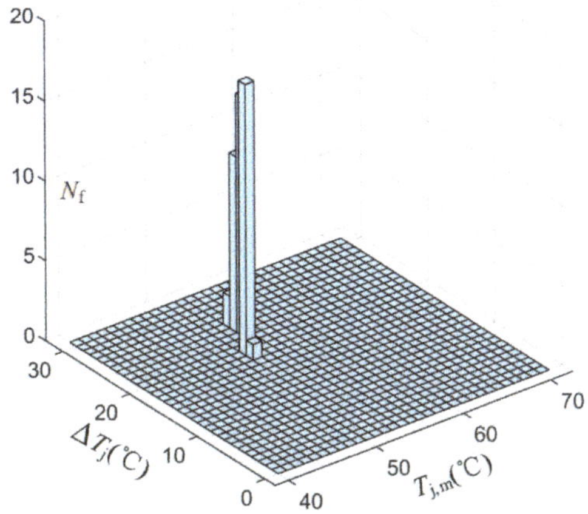

Figure 15. Thermal load histogram of the SiC MOSFET under test before control.

The LESIT model considers the relationship between the number of device failure cycles and the average cycle junction temperature and the amplitude of the cycle junction temperature change. The life prediction model is as follows [4,5]:

$$N_f = A\Delta T_j^{\alpha} e^{\frac{E_\alpha}{k_B(T_{j,m}+273)}} \tag{8}$$

In Formula (8), N_f is the number of device failure cycles. A and α are the experimental coefficients, k_B is the Boltzmann constant, and E_α is the activation energy.

The parameters A and α in Formula (8) are determined according to the power cycle test. In the power cycle experiment, when the time ton15 s of the input stress is applied, the stress is applied to the solder layer. The results of these tests provide data on the reliability of the equipment under the test under specific operating conditions, which is usually provided by the manufacturer. The values of the data are shown in Formula (9) [19].

$$A = 3 \times 10^5, \alpha = -5.039, E_\alpha = 9.98 \times 10^{-20}, k_B = 1.38 \times 10^{-23}. \tag{9}$$

The life estimation is based on Miner's criterion or damage accumulation theory. In practical applications, by correlating the ratio of the number of thermal cycles N_v and the number of relative device failure cycles N_f with ΔT_j and $T_{j.\ m}$, the damage is associated with each stress state. It is assumed that the damage caused by the different temperature fluctuations is independent of each other. So, the total damage Q can be obtained by adding all the damage [20]. Miner's linear fatigue accumulation theorem holds that the damage caused by each cycle is constant regardless of the amplitude of the stress cycle. and these damages can be simply added. This assumption simplifies the prediction process of fatigue life, allowing engineers to easily estimate the life of the material in actual use based on the fatigue performance test results of the material. The linear cumulative damage theory of Palmgren–Miner, called Miner's rule, is reasonable under certain conditions which can simplify the prediction process of fatigue life and give relatively accurate results.

$$Q = \sum_{i=1}^{k} Q_{v,i} = \sum_{i=1}^{k} \frac{N_{v,i}}{N_{f,i}} \tag{10}$$

$$T = \frac{1}{Q} \tag{11}$$

Figure 15 shows the amplitude, fluctuation times, and average value of the junction temperature of a device before the conduction loss control of the body diode can be obtained. Through Figure 16, the amplitude, fluctuation times, and average value of the junction temperature of the device after the body diode conduction loss control can be obtained. By calculating the $Q_{v,i}$ of all the i th operating points of the electric drive inverter, the overall damage of the SiC power device can be predicted. If the damage is less than 1, the power device can withstand the thermal stress in the entire task curve. Otherwise (if the damage is greater than or equal to 1), the equipment is considered to be faulty. T expresses how many cycles the device under the test can cycle before failure occurs. For the 'cycle', the stress input of the rain flow must be considered. Assuming that a cycle corresponds to one second of the steady-state junction temperature, the device life formula can be expressed as follows:

$$Lifetime = \sum_{i=1}^{k} \frac{1}{360 * h\omega Q * 365} \tag{12}$$

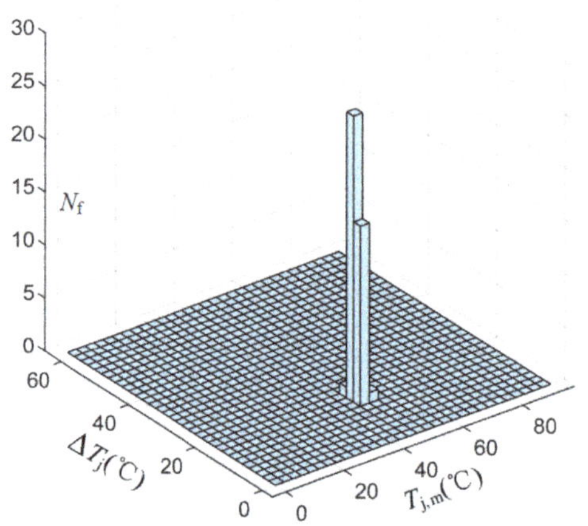

Figure 16. Thermal load histogram of the SiC MOSFET under test after control.

In the formula, ω is the ratio of the i th 'cycle' times to the total 'cycle' times, h is the daily working time of the device, and the value is 24. The Q_A and Q_B of the device before and after the body diode conduction loss control can be obtained by Formulas (8), (10) and (12), and the life of the device before and after the body diode conduction loss control Lifetime_A and Lifetime_B can be obtained by Formula (12).

Table 6 shows the results obtained by using linear fatigue cumulative damage theory and life model. Because the actual working conditions of the electric drive inverter are much more complex than the working conditions listed in this paper, the actual life will be shorter than the life calculated in this paper.

Table 6. Theoretical calculation results of linear fatigue cumulative damage.

	Cumulative Damage	Device Lifetime
Before control	1.45×10^{-8}	21.87 years
After control	1.18×10^{-8}	26.87 years

The calculation results of linear fatigue cumulative damage can prove that under the control of body diode conduction loss, the cumulative damage of the device is reduced from 1.45×10^{-8} to 1.18×10^{-8} and the life of the device after junction temperature control is 1.23 times that before junction temperature control. Experiments show that the proposed method of adjusting the junction temperature of the device through the conduction loss of the body diode can improve the life of the device and improve its operational reliability.

5. Conclusions

Aiming at the method of improving the reliability of SiC MOSFET devices in converters, the method of the smooth junction temperature control of SiC devices based on the adjustment of the body diode conduction loss of SiC MOSFET devices is proposed. The body diode conduction principle of the SiC MOSFET device was analyzed. The feasibility and working principle of adjusting the junction temperature using body diode conduction loss were expounded. The influence of body diode conduction time on a converter was expressed by formula derivation. The correctness of the proposed body diode conduction loss adjustment was further verified using a Saber simulation. Finally, the proposed body diode conduction loss adjustment was verified by experiments. The rain flow counting method and the linear fatigue damage formula were introduced. The life of the device before and after the body diode conduction loss control was evaluated and compared by the rain flow counting method and the linear fatigue damage formula. According to the research methods of this paper, the main conclusions are as follows:

1. Based on the characteristics of SiC MOSFET devices, an active thermal management method based on body diode conduction loss was proposed and the correctness of the proposed method was verified via a simulation and an experiment. Under the comparison of the calculation results of the life prediction model, the conduction loss adjustment of a body diode can effectively extend the life of a SiC device and improve its operational reliability;
2. The control of the body diode conduction time is realized by controlling the dead time of the converter. When the dead time of the converter is too large, it will affect the distortion of the output waveform of the converter.

The method proposed in this paper is based on the conduction loss adjustment of the body diode of the SiC MOSFET device to realize the smooth control of the junction temperature of the SiC device. This method is relatively independent and suitable for most converters. The control method of this method is simple, and the dead time adjustment can be realized by hardware or software.

However, this method is suitable for devices with a large body diode conduction voltage drop; otherwise, the temperature regulation effect is not good. Under actual operating conditions, in order to improve the reliability of the converter, diodes are generally connected in parallel at both ends of the SiC MOSFET device to limit the use of the SiC MOSFET device body diode. The selection of this aspect can be carried out in subsequent research.

Author Contributions: J.W.: conceptualization, methodology, visualization, and writing—original draft. Y.W. (Yunpeng Wei): resources, writing—review and editing, and supervision. Y.W. (Yuntao Wu): software and validation. Z.W.: formal analysis and investigation. X.L.: visualization and writing—review and editing. X.W.: data curation and supervision. All authors have read and agreed to the published version of the manuscript.

Funding: This work was supported by the Innovation Project of GUET Graduate Education (No. 2023YCXS020) and in part by the Guangxi Science and Technology Plan Project under Grant Guike AD20297046.

Data Availability Statement: Data will be made available on request.

Conflicts of Interest: The authors declare no conflicts of interest.

References

1. Minghui, C.; Chunning, W.; Hao, W. Application prospect of SiC power semiconductor devices in modern systems. *Telecom Power Technol.* **2018**, *35*, 11–13.
2. Zhang, J.; Shen, H.; Du, X.; Chen, R. Condition Monitoring the Inhomogeneous Thermal Fatigue of Multichip IGBT Module Based on the Thermal Attenuation Coefficient. *IEEE Trans. Power Electron.* **2025**, *40*, 2114–2125. [CrossRef]
3. Wang, B.; Zhou, L.; Zhang, Y.; Wang, K.; Du, X.; Sun, P. Active Junction Temperature Control of IGBT Based on Adjusting the Turn-off Trajectory. *IEEE Trans. Power Electron.* **2018**, *33*, 5811–5823. [CrossRef]
4. Tang, X.; Zhang, Y.; Sai, Z.; Han, R.; Wu, J.; Pan, Y. Influence of Stray Capacitance on Dynamic Test Results of Power Semiconductor devices. In Proceedings of the 2019 4th IEEE Workshop on the Electronic Grid (eGRID), Xiamen, China, 11–14 November 2019; pp. 1–4.
5. Held, M.; Jacob, P.; Nicoletti, G.; Scacco, P.; Poech, M.H. Fast power cycling test of IGBT modules in traction application. In Proceedings of the Second International Conference on Power Electronics and Drive Systems, Singapore, 26–29 May 1997; Volume 1, pp. 425–430.
6. Wang, X.; Castellazzi, A.; Zanchetta, P. Temperature control for reduced thermal cycling of power devices. In Proceedings of the 2013 15th European Conference on Power Electronics and Applications (EPE), Lille, France, 2–6 September 2013; pp. 1–10.
7. Guo, Y. Study on IGBT Switch Trajectory Adjustment and Junction Temperature Control Based on Active Gate Drive Control. Master's Thesis, South China University of Technology, Guangzhou, China, 2020.
8. Sun, Y.; Sun, L.; Esmaeli, A.; Zhao, K. A Novel Three Stage Drive Circuit for IGBT 2006. In Proceedings of the 1st IEEE Conference on Industrial Electronics and Applications, Singapore, 24–26 May 2006; pp. 1–6.
9. Luo, H.; Iannuzzo, F.; Ma, K.; Blaabjerg, F.; Li, W.; He, X. Active gate driving method for reliability improvement of IGBTs via junction temperature swing reduction. In Proceedings of the 2016 IEEE 7th International Symposium on Power Electronics for Distributed Generation Systems (PEDG), Vancouver, BC, Canada, 27–30 June 2016; pp. 1–7.
10. Falck, J.; Andresen, M.; Liserre, M. Active thermal control of IGBT power electronic converters. In Proceedings of the IECON 2015—41st Annual Conference of the IEEE Industrial Electronics Society, Yokohama, Japan, 9–12 November 2015; pp. 000001–000006.
11. Trzynadlowski, A.M.; Kirlin, R.L.; Legowski, S.F. Space vector PWM technique with minimum switching losses and a variable pulse rate for VSI. *IEEE Trans. Ind. Electron.* **1997**, *44*, 173–181. [CrossRef]
12. Du, X.; Li, G.; Sun, P.; Zhou, L.; Tai, H.M. A hybrid modulation method for lifetime extension of power semiconductors in wind power converters. In Proceedings of the 2015 IEEE Applied Power Electronics Conference and Exposition (APEC), Charlotte, NC, USA, 5–19 March 2015; pp. 2565–2570.
13. Yunhai, W.; Minyou, C.; Wei, L. Review on active thermal control methods based on junction temperature swing smooth control of IGBTs. *Trans. China Electrotech. Soc.* **2022**, *37*, 1415–1430.
14. Ozkan, G.; Papari, B.; Hoang, P.H.; Deb, N.; Edrington, C.S. An Active Thermal Control Method for AC-DC Power Converter with Sequence-based Control Approach. In Proceedings of the 2019 IEEE Electric Ship Technologies Symposium (ESTS), Washington, DC, USA, 14–16 August 2019; pp. 263–267.
15. Falck, J.; Buticchi, G.; Liserre, M. Thermal Stress Based Model Predictive Control of Electric Drives. *IEEE Trans. Ind. Appl.* **2016**, *54*, 1513–1522. [CrossRef]
16. Wu, J.; Zhou, L.; Wang, B. Internal thermal management of power converter based on switching trace adjustment. *J. Power Supply* **2016**, *14*, 46–52.
17. Xie, J. Study on Temperature Characteristics of High Votage SiC MOSFET Body Diode. Master's Thesis, North China Electric Power University, Beijing, China, 2022.
18. Jianfeng, Z. Analysis of the impact of dead-time in inverters on output voltage. *Power Electron.* **2007**, *2*, 792–797.
19. Lai, W.; Chen, M.; Ran, L.; Alatise, O.; Xu, S.; Mawby, P. Low ΔTj Stress Cycle Effect in IGBT Power Module Die-Attach Lifetime Modeling. *IEEE Trans. Power Electron.* **2016**, *31*, 6575–6585. [CrossRef]
20. Wölfle, J.; Röser, T.; Nitzsche, M.; Tröster, N.; Stempfle, M.; Ruthardt, J.; Roth-Stielow, J. Model based temperature control system to increase the expected lifetime of IGBT power modules executed on a neutral point diode clamped three level inverter. In Proceedings of the 2017 19th European Conference on Power Electronics and Applications (EPE'17 ECCE Europe), Warsaw, Poland, 11–14 September 2017; pp. 1–9.

energies

MDPI

Article

Online Monitoring Method for Opening and Closing Time of 10 kV Spring Energy Storage Circuit Breaker Based on Transient Electrical Signal Characteristic Point Marking and Self-Calibration

Liting Weng [1], Jiangtao Xiao [1], Zhaochuang Zhang [1], Jingang Wang [2], Yuchuan Wen [3], Fan Zhang [2], Xingyu Zhang [2], Lingyi Ma [2] and Pengcheng Zhao [2,*]

1 Xiluodu Hydropower Plant, Zhaotong 657300, China; weng_liting@ctg.com.cn (L.W.); xiao_jiangtao@ctg.com.cn (J.X.); zhang_zhaochuang@ctg.com.cn (Z.Z.)
2 State Key Laboratory of Power Transmission Equipment Technology, School of Electrical Engineering, Chongqing University, Chongqing 400044, China; jingang@cqu.edu.cn (J.W.); zhangfan023@stu.cqu.edu.cn (F.Z.); 202311131307@stu.cqu.edu.cn (X.Z.); 202311131275@stu.cqu.edu.cn (L.M.)
3 Three Gorges Ecological Environment Co., Ltd., Zhaotong 657300, China; wen_yuchuan@ctg.com.cn
* Correspondence: zhaopengcheng@cqu.edu.cn

Abstract: Among all circuit breaker faults, mechanical failures account for a considerable proportion, and online monitoring of their mechanical characteristics is of great practical significance. The opening and closing time is a very important feature of the mechanical characteristics of the circuit breaker. Online monitoring of the opening and closing time of the circuit breaker has always been the focus and difficulty of the intelligent technology of switchgear. In this paper, for a 10 kV spring energy storage vacuum circuit breaker, transient voltage and current signals are innovatively used to calibrate the opening time, breaking time, and closing time, and an online monitoring method for the opening and closing time of a vacuum circuit breaker based on transient electrical signals is proposed. An online monitoring platform was built and a multi-group closing test was carried out to simulate the power plant environment. The opening and closing time samples of a spring energy storage vacuum circuit breaker were measured and compared with the measurement results of the mechanical properties tester. The comparison results show that this method has good stability, and the calculation error is controlled within 1% after self-calibration, which provides a new idea for the online monitoring research of the mechanical characteristics of spring energy storage vacuum circuit breakers.

Keywords: circuit breaker; online monitoring; opening and closing time; self-calibration

1. Introduction

In modern power systems, high-voltage circuit breakers are one of the most important control and protection devices. In order to prevent high-voltage circuit breakers from failure, they need to be frequently inspected and maintained. Too low a frequency of circuit breaker inspection will reduce system reliability, while too frequent inspections will bring new man-made risks and higher costs. Among all circuit breaker failures, mechanical failures account for a considerable proportion [1]. Therefore, online monitoring of their mechanical characteristics is of great practical significance.

The opening and closing time of the circuit breaker is a very important parameter in the online monitoring of the circuit breaker and an important indicator to measure the performance of the circuit breaker. The opening and closing time is closely related to the health of the operating mechanism and directly affects the breaking performance of the circuit breaker [2]. Therefore, accurately extracting the opening and closing time characteristic values through the online monitoring system is of great significance for judging the health and working status of the circuit breaker.

Regarding the online monitoring method of the mechanical characteristics of the vacuum circuit breaker during the opening and closing process, some scholars have conducted research on the aspects of the opening and closing coil current signal [3,4], the moving contact travel–time curve during the opening and closing operation [5], and the circuit breaker vibration signal [6–12]. Reference [13] proposed an online monitoring method for the closing point based on the travel curve of a vacuum circuit breaker. It analyzed the stress conditions of the moving contact before and after the closing point, and based on the analysis results, proposed using the closing speed curve as key information. The wavelet decomposition and reconstruction method was used to extract feature quantities from the closing speed curve for calculating the closing point. Reference [14] extracted time characteristics from the opening/closing coil current waveform signal and the circuit breaker auxiliary contact voltage waveform signal, constructed a degradation degree index that takes into account the influence of monitoring data uncertainty, and introduced a sequential Bayesian algorithm to achieve real-time updating of circuit breaker status assessment results that integrate historical records and monitoring data; ref. [15] proposed a circuit breaker state assessment method based on vibration signal envelope analysis; and [16] proposed a novel, non-intrusive method for circuit breaker arcing time, which combined vibration-acoustic fusion and convolutional neural network [17]. By measuring and analyzing the high-definition transient voltage waveform of the high-voltage circuit breaker closing process, the time interval between the auxiliary contact bounce and arc extinguishing is extracted as the characteristic parameter to realize the state evaluation of the auxiliary contact bounce. Reference [18] selected the closing coil current and travel signal as the experimental monitoring quantities, extracted three types of principal component features with orthogonal contributions, and constructed a ternary feature map based on this, which intuitively displayed the evolution path of the curve under different faults. Reference [19] normalized and preprocessed the different characteristics of the circuit breaker from multiple angles such as current and vibration signals, used the PCA method to fuse the features and extract the principal components containing the richest information on mechanical property degradation, and obtained a comprehensive health index of the mechanical properties of the energy storage operating mechanism. References [20,21] used the method of angular displacement sensor to detect the mechanical characteristics of the circuit breaker. Based on the analysis of the circuit breaker connecting rod structure, the relationship between the main shaft rotation angle and the contact line displacement was established, and the accuracy of the angular displacement sensor for contact monitoring was verified through experiments.

The traditional method of mechanical and time characteristic testing must be carried out under power outage maintenance. The power outage offline switch characteristic test will reduce the number of switch operations and reduce the switch operation life. The switch wiring is also prone to component damage. In addition, the offline mechanical characteristic test of the switch consumes human resources and cannot continuously measure data, which is not conducive to the fault diagnosis and status monitoring of the circuit breaker. This paper takes the ZN28-12 indoor vacuum spring energy storage circuit breaker commonly used in power plants as the research object, and studies the method for determining the opening and closing time during its online operation. An online monitoring method for the opening and closing time of the circuit breaker that only relies on voltage and current sensors is proposed. This method starts from the characteristic that the transient waveform of the electrical signal before and after the opening/closing point changes significantly, and innovatively uses the primary side transient signal to calibrate the opening time, breaking time, and closing time. Compared with the traditional calibration of the secondary side opening and closing coil signal or displacement sensor and vibration sensor, this method is more direct. This use of electrical signals to realize online monitoring of mechanical characteristics can more truly and effectively reflect the operating status of the circuit breaker, with a faster response speed, and has guiding significance for real-time detection of circuit breaker fault information.

2. Calibration of Opening and Closing Points Based on Transient Electrical Signal Characteristics of Circuit Breaker Opening and Closing

The transient voltage and current during the opening and closing process of the circuit breaker are caused by the arc generated between the moving and static contacts during the opening and closing process. The "arcing" phenomenon occurs when the circuit breaker is opened, and the "pre-breakdown" phenomenon occurs when the circuit breaker is closed. "Arcing" and "pre-breakdown" are essentially the breakdown of the vacuum medium at the break when the moving and static contacts are not fully in contact. According to the classical vacuum break breakdown theory, when the voltage applied to the break exceeds the insulation strength of the gap medium, the break will break down, causing transient electrical signals to appear.

Through Paschen's law or Paschen curve, we know that when the voltage intensity of the vacuum circuit breaker is greater than the insulation strength of the vacuum, a breakdown will occur, and an arc will be generated. When the circuit breaker is opening and closing, the distance between the moving and static contacts will change, so the insulation recovery strength between the moving and static contacts will show dynamic changes. When the voltage intensity between the contacts is greater than the recovery strength of the insulation, the medium will be broken down and an arc will be generated. The generation of the arc is accompanied by transient voltage and current.

Due to the existence of the cut-off value, the transient current signal generally exists for a shorter time than the transient voltage signal, because when the current flowing through the circuit breaker is less than the cut-off value, the transient current signal disappears. However, at this time, a transient recovery voltage will be generated between the contacts, which is part of the transient voltage existence process. Comprehensive calibration can be performed through transient voltage and current signals. For example, the starting points of transient voltage and transient current correspond to the time when the arc is generated. In some tests where the transient voltage is not obvious, the transient current can assist in calibrating the characteristic points. It can be seen that there is a close connection between the distance (mechanical stroke) between the moving and static contacts of the circuit breaker, the arc, and the transient electrical signal. After the circuit breaker contacts are separated, an arc is generated between the contacts, and the arc is extinguished when the arc current passes through zero.

2.1. Relationship Between Transient Electrical Signal of Circuit Breaker Opening and Closing and Moving Contact Travel Curve

When the circuit breaker receives the opening signal, it starts to perform the opening operation. At the moment when the moving and static contacts of the circuit breaker separate, the distance between the contacts is the shortest. If the voltage reaches the breakdown voltage of the vacuum, breakdown occurs, an arc is generated, and a transient signal is generated at the same time. When the distance between the moving and static contacts increases and is not enough to cause the vacuum to break down, the arc is extinguished, and the transient signal disappears. Figure 1 is a timing diagram of the opening transient electrical signal and the moving contact travel curve. It can be seen from the figure that after the circuit breaker receives the opening command at time T_{command}, it takes a time interval T_{opening} from the energization of the circuit breaker opening coil to the separation of the circuit breaker contacts.

$$T_{\text{opening}} = T_{\text{seperate}} - T_{\text{command}} \tag{1}$$

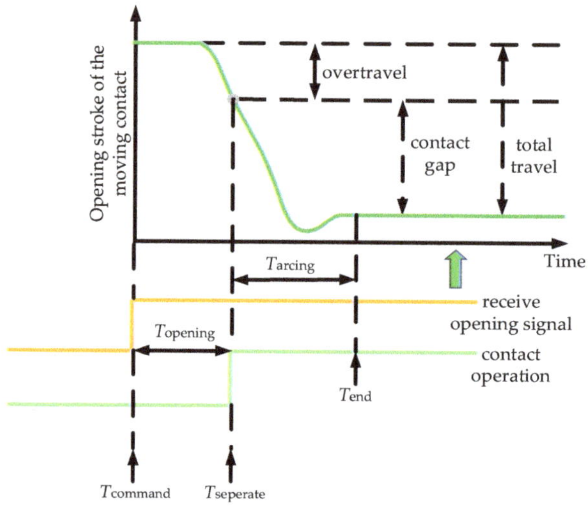

Figure 1. Timing diagram of the opening transient electrical signal and the moving contact travel curve.

$$T_{\text{arcing}} = T_{\text{end}} - T_{\text{seperate}} \quad (2)$$

The time point of contact action is T_{seperate}. After the contact action, since the contact surface of the circuit breaker cannot be completely flat, the current will always flow through the separated convex point when the moving and static contacts separate, causing local overheating and melting, and then generating a diffuse arc. The time from the arc generation to the extinction is called the arcing time. Transient voltage and current are also generated with the arcing, and their existence time is T_{arcing}.

The arcing time is not fixed, but it has certain random characteristics. The arcing time of each circuit breaker satisfies the normal distribution law, and the existence time of transient electrical signals (voltage, current) also satisfies certain normal distribution laws. Through multiple groups of tests, the distribution interval of the transient electrical signal existence time T_{arcing} can be measured.

Similar to the opening operation, after receiving the closing command, it takes a period of time ΔT_{known} from the energization of the circuit breaker tripping coil to the start of the circuit breaker contact action. After the contact moves, as the gap between the breaker and the circuit breaker continues to decrease, pre-breakdown may occur under certain conditions. The occurrence of pre-breakdown will cause the circuit to be connected in advance before the contacts touch. As shown in Figure 2, the timing diagram of the closing transient electrical signal and the moving contact travel curve, the time of the entire closing process is divided into three parts, as shown in the formula:

$$T_{closing} = \Delta T_{\text{known}} + \Delta T_{stochastic} + T_{\text{prestrike}} \quad (3)$$

where T_{command} is the moment when the circuit breaker receives the closing command, and T_{seperate} is the moment when the moving contact starts to move. Pre-breakdown occurs after the moving contact starts to move, $\Delta T_{\text{stochastic}}$, and its value is random but satisfies the normal distribution law. The arcing time after pre-breakdown occurs is $T_{\text{prestrike}}$, and the existence time of the transient electrical signal in the closing process corresponds to its arcing time. When the closing operation is performed, it can be mainly divided into the following stages: the circuit breaker receives the closing signal, the closing coil is energized, and the iron core begins to move until the moving and static contacts of the circuit breaker are close to each other, generating transient voltage and current signals; the time from the transient electrical signal generated between the moving and static contacts of the circuit breaker to the time when the moving and static contacts of the circuit breaker just begin to touch each other is the arcing time, and the voltage signal in the system is the transient signal; from the time when the moving and static contacts of the circuit breaker just begin to touch each other to the time when the circuit breaker is closed, the transient voltage and current in the system disappear.

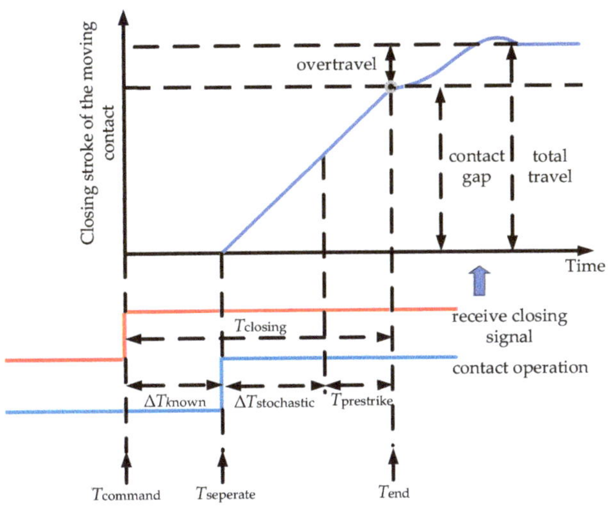

Figure 2. Timing diagram of closing transient electrical signal and moving contact travel curve.

2.2. Calibration of Opening and Closing Characteristic Points Based on Transient Electrical Signals

Figure 3 shows the waveform and time point calibration of the transient electrical signal (voltage, current) of the circuit breaker when closing. According to the principle of characteristic point identification, time T_2 is the moment when the transient electrical signal appears, which produces the "pre-breakdown" phenomenon; time T_3 is the moment when the transient electrical signal disappears, corresponding to the just-closed point in the moving contact travel–time curve, and is also the calibration point for calculating the closing time of the circuit breaker. By measuring the closing coil pulse signal, it can be known that the closing signal occurs at time T_1, the arcing time of closing $T_{prestrike}$ depends on the existence time of the transient electrical signal, and the just-closed time $T_{closing}$ depends on the time difference between the disappearance point of the transient electrical signal and the occurrence point of the closing signal.

$$T_{prestrike} = T_3 - T_2 \tag{4}$$

$$T_{closing} = T_3 - T_1 \tag{5}$$

Figure 4 shows the waveform and time point calibration of the transient electrical signal (voltage, current) of the circuit breaker during opening. According to the principle of characteristic point identification, the moment T_5 is the moment when the transient electrical signal appears, which corresponds to the just-opening point in the moving contact travel time curve, and is also the calibration point for calculating the circuit breaker opening time; the moment T_6 is the moment when the transient electrical signal disappears, and the arcing phenomenon will also disappear at this time. By measuring the pulse signal of the opening coil, it can be known that the opening signal occurs at T_4, the arcing time of the T_{arcing} of the opening depends on the existence time of the transient electrical signal, and the just-opening time T_{opening} depends on the time difference between the appearance point of the transient electrical signal and the occurrence point of the opening signal.

$$T_{\text{arcing}} = T_6 - T_5 \tag{6}$$

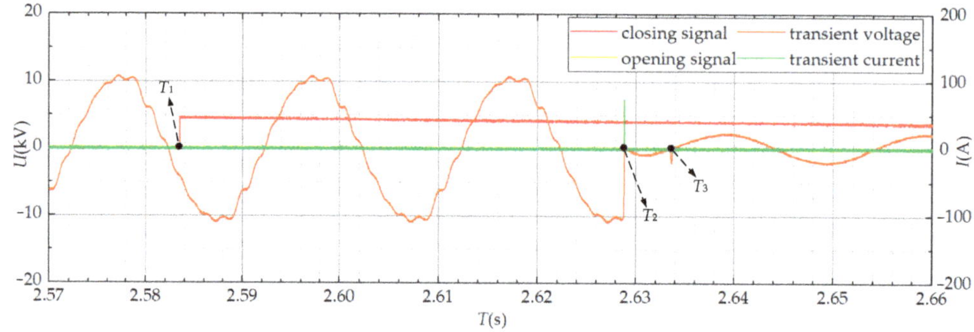

Figure 3. Waveform and time point calibration of transient electrical signals (voltage, current) when closing the circuit breaker.

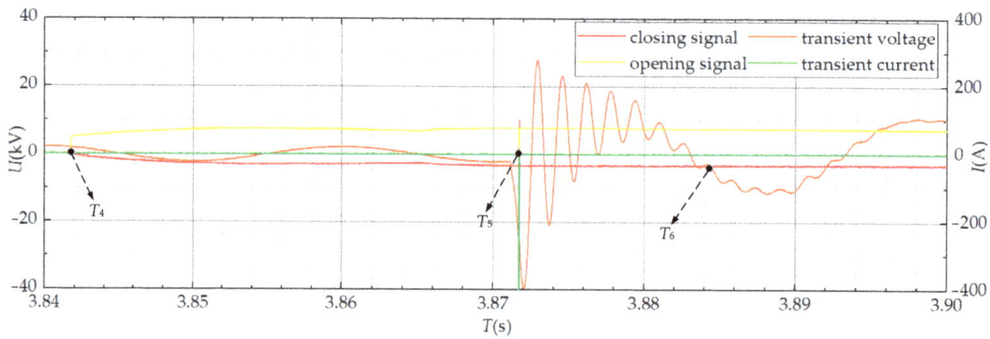

Figure 4. Waveform and time point calibration of transient electrical signals (voltage, current) during opening.

$$T_{\text{opening}} = T_5 - T_4 \tag{7}$$

As can be seen from Figure 3 and Table 1, when performing a closing operation, it can be mainly divided into the following stages:

- From the time when the circuit breaker receives the closing signal, and the closing coil is energized to drive the iron core to start moving until the moving and static contacts of the circuit breaker are close to each other and an arc is generated, the voltage and current signals in the system are both industrial frequency electrical signals.
- From the time when an arc is generated between the moving and static contacts of the circuit breaker to the time when the moving and static contacts of the circuit breaker just begin to touch each other is the arc burning time, and the voltage and current signals in the system are transient signals. Through the analysis in the previous stage, the time point needs to be calibrated by current assistance.
- From the time when the moving and static contacts of the circuit breaker just begin to touch each other to the time when the circuit breaker is closed, the voltage and current signals in the system are industrial frequency electrical signals.

Table 1. Characteristic quantity of electrical signal during opening and closing operation.

Operate	Time	Feature Quantity
Opening	Opening trigger—just opened	Power frequency electrical signal
	Arcing time	Transient electrical signal
	Arcing ends—opening completed	Power frequency electrical signal
Closing	Closing trigger-arcing time	Transient electrical signal
	Arcing time—just closed	Power frequency electrical signal
	Just closed-closing completed	Transient electrical signal

As can be seen in Figure 4 and Table 1, when the opening operation is performed, it can be mainly divided into the following stages:

- From the time when the circuit breaker receives the opening signal, the opening coil is energized to drive the iron core to start moving until the moving and static contacts of the circuit breaker begin to separate, and the voltage signal and current signal in the system are both industrial frequency signals.
- From the time when the moving and static contacts of the circuit breaker begin to separate to the time when the arc of the moving and static contacts no longer breaks through is the arcing time, the voltage signal and current signal in the system are transient signals, and the time point can be directly calibrated by the voltage signal.
- From the time when the arcing of the moving and static contacts of the circuit breaker ends to the time when the circuit breaker is opened, the voltage signal and current signal in the system are industrial frequency electrical signals.

3. Design of Online Monitoring System for Circuit Breaker Opening and Closing Time Based on Voltage and Current Transient Signals

The overall design scheme of the online monitoring system for transient electrical signals at the circuit breaker opening and closing is shown in Figure 5. The capacitive voltage sensor converts the transient voltage at the break into a voltage signal in real time; the Rogowski coil measures the electromagnetic signal at the break, converts it into a current signal through an integrator, and then uses an acquisition card to convert the transient electrical signal into a digital signal and collect it. The host computer software processes and stores the collected signals, scans the slope of the waveform, and identifies the time point when the transient electrical signal is generated or disappears. Finally, it can realize the real-time display of the electrical signal waveform and the automatic calculation and storage of the opening and closing time.

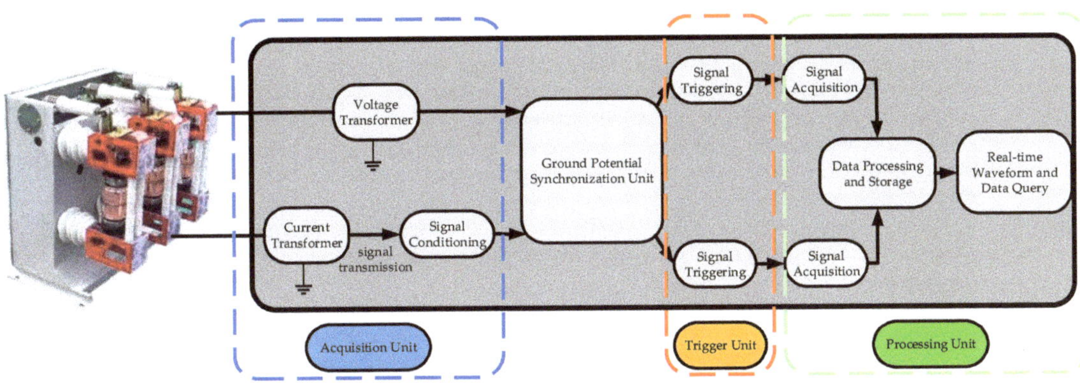

Figure 5. Online monitoring system architecture diagram.

The realization of online monitoring first requires transient electrical signal data acquisition. It is necessary to collect the transient original electrical signals at both ends of the circuit breaker break. The transient voltage signal is collected by a capacitive voltage sensor, and the transient current signal is collected by a Rogowski coil with an integrator. Finally, the acquisition card stores the data. The process of the acquisition program is shown in Figure 6.

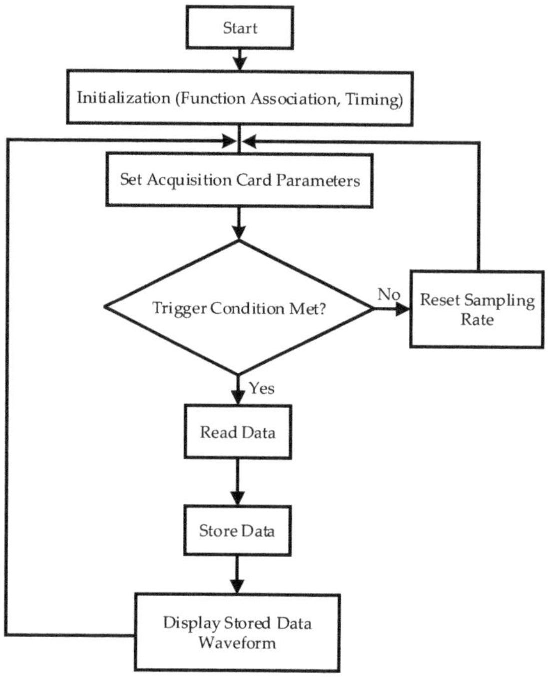

Figure 6. Signal acquisition flow chart.

The first step of program startup is to initialize, and then set the relevant acquisition parameters, such as sampling rate, trigger threshold, channel range, input impedance, coupling mode, etc. The trigger threshold is set to prevent interference from clutter. Selecting an appropriate threshold is the premise for ensuring that the signal is not lost and filtering. When the acquisition signal reaches the trigger condition, the data will be read and stored by the host computer software, and the acquired transient electrical signal waveform will be displayed in real time on the display interface of the host computer. If the trigger condition is not met, the acquisition parameters need to be reset, considering whether the sampling rate meets the transient electrical signal acquisition bandwidth requirements of the current system.

After filtering the collected transient voltage and current waveforms of opening and closing, the clutter in the power grid is filtered out. By performing slope scanning on the filtered waveform, the time points when the transient electrical signal appears and disappears are obtained. When there is no transient signal, the slope of each point in the waveform corresponds to the sine curve, but when the transient signal appears, the slope no longer satisfies the corresponding relationship. Therefore, the characteristic points of the transient electrical signal can be found by the slope scanning method. Then, the calibration method specified in this article is used to calibrate the opening and closing points, and finally, the opening and closing time is automatically calculated.

4. Self-Calibration of Opening and Closing Time Measurement Based on Transient Electrical Signals

4.1. Selection of Reference Value for Self-Calibration of Opening and Closing Time

The proposed method is compared and corrected with the circuit breaker mechanical characteristic tester, taking advantage of the circuit breaker mechanical characteristic tester's stable measurement and applicability to multiple scenarios. In the same system, the measurement value of the circuit breaker mechanical characteristic tester is used as a benchmark to calibrate the error of the online monitoring system based on transient electrical signals. The wiring diagram of the mechanical tester used in this experiment is shown in Figure 7.

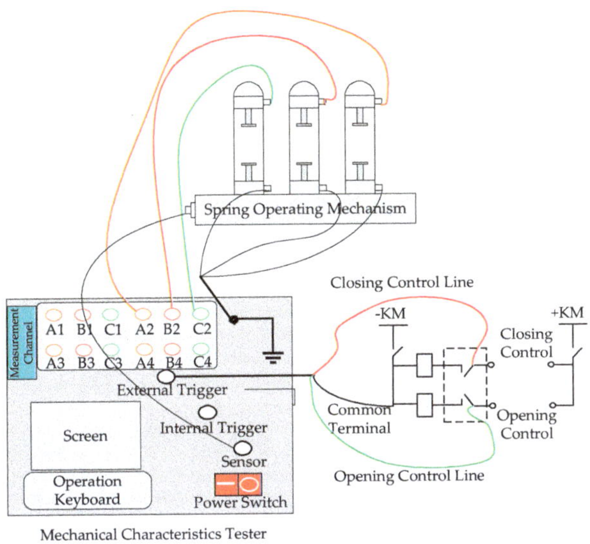

Figure 7. Circuit breaker mechanical characteristics tester test wiring diagram.

The mechanical characteristic test of the vacuum circuit breaker is completed by a digital mechanical characteristic tester. In the process of online monitoring, in order to ensure safety, the background control usually pressurizes the opening and closing coils through an external power supply. When wiring, the auxiliary contacts of the circuit breaker need to be connected to prevent the coil from burning due to long-term power supply. The time signal is drawn from the upper and lower plum blossom contacts of the circuit breaker.

When the circuit breaker is in the open state, the closing test is performed with the closing signal received as the starting point of the timing, and the timing is stopped when the upper and lower contacts are metal-conducting. This period of time is the closing time. When the circuit breaker is in the closed state, the opening test is performed with the opening signal received as the starting point of the timing, and the timing is stopped when the upper and lower contacts are separated. This period of time is the opening time.

4.2. Self-Calibration Method for Opening and Closing Points

The method adopted in this paper is to calibrate the opening and closing time points of the circuit breaker through transient electrical signals, so as to calculate the opening and closing time. Due to the influence of noise and other factors, there will be certain interference in the acquisition of electrical signals. In order to improve the accuracy of this method, an error compensation link is introduced in the calculation process. Most of the research on error compensation of mechanical and time characteristic parameters of circuit breakers is still at the level of error source modeling, and most of them are offline compensation methods. Based on this, this paper gives an intelligent model-free real-time error compensation method that can effectively process data online, combined with preliminary error mean compensation and further compensation of intelligent compensation module to ensure that the final error meets the expected range. The compensation process is shown in Figure 8.

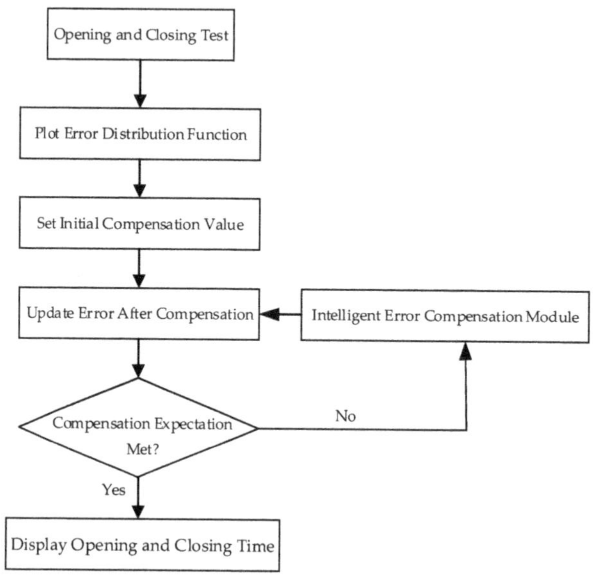

Figure 8. Compensation Flowchart.

A 10 kV opening and closing test platform was built to perform opening and closing operations, and transient electrical signal waveforms were collected. The opening and closing calculation time was obtained by calibrating the opening and closing points. At the same time, the advantage of the circuit breaker mechanical characteristic tester was the stability of the measurement, and the opening and closing time was measured as a benchmark to obtain the calculation error value of the opening and closing time point based on the transient electrical signal. After multiple groups of tests, the error normal distribution curve was obtained, and the error mean was used as the initial compensation amount. After compensation, it was compared with the benchmark value again, and the intelligent error compensation module was used to compensate until the expected value was met.

The intelligent error compensation module adopts the model-free iterative learning error compensation method, and its structural block diagram is shown in Figure 9. After obtaining the error signal, the LMS (Least Mean Squares) algorithm is used. The LMS algorithm dynamically adjusts the compensation coefficient by minimizing the mean square error between the output signal and the expected signal, so that it can adapt to changes in real time under different environments. After a large number of opening and closing tests, a learning law is obtained to compensate for the error signal. The memory stores historical data of past errors and compensations, and the compensation required for each opening and closing time can be obtained through learning.

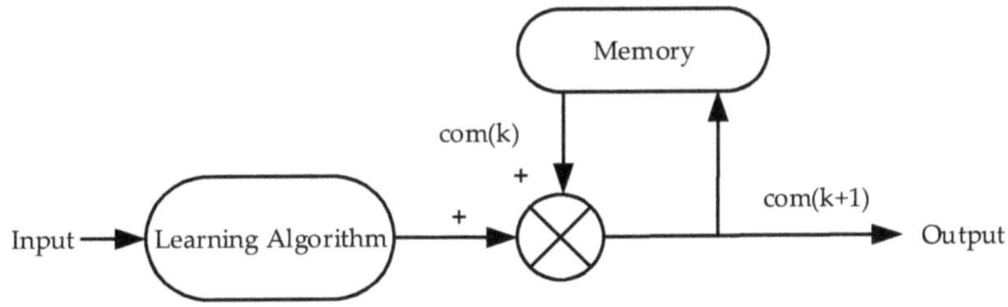

Figure 9. Error compensation iterative learning flow chart.

Assume that the actual measured opening and closing time is t_m, the reference time is t_r, the time error is e_t, the mean error is μ_e, the intelligent compensation amount is τ, and the error is less than the expected value of e_r.

Read the real-time data t_m, and after compensation by the mean error μ_e, the initial compensation opening and closing time t_{c1} can be obtained as:

$$t_{c1} = t_m + \mu_e \tag{8}$$

The initial time error $e_t{}'$ is:

$$e_t{}' = t_r - t_{c1} \tag{9}$$

Check whether the error is within the expected range. If so, output t_{c1}. Otherwise, call the intelligent compensation module to compensate for the error. The intelligent compensation amount is τ:

$$\tau = f(\text{Mean Squared Error}, e_t{}') \tag{10}$$

Calculate the time after final compensation:

$$t_{c2} = t_{c1} + \tau \tag{11}$$

The current measurement and compensation results are stored in the database for the intelligent compensation module to learn and optimize the compensation function.

5. Case Study

This paper takes the ZN28-12 indoor 10 kV vacuum spring energy storage circuit breaker commonly used in power plants as the research object, as shown in Figure 10. An online monitoring system for circuit breaker opening and closing is built in the laboratory, as shown in Figure 11.

The voltage sensor used in the online monitoring system is a TCF10-300 low-damping high-voltage capacitive voltage divider. After encapsulation, its external insulation is cast with epoxy resin. The overall appearance is similar to that of a lightning arrester. The rated voltage is 10 kV, the voltage division ratio is 1000:1, the sampling rate is 10 MHz, and the accuracy is ±0.5%. The current transformer used in the system is a CWT-30B/4/700 Rogowski coil, in which the lead wire is 4 m and the ring circumference is 700 mm. The lead wire length and the ring circumference of the coil have a certain margin, which has good field adaptability, a sampling rate of 10 MHz, and an accuracy of ±0.2%. The data acquisition card of this system supports 8-channel synchronous data sampling. While sampling the three-phase current and three-phase voltage, it also leaves a margin of 2 channels for other needs in the later stage. Its sampling rate can reach up to 80 MHz.

Capacitive voltage sensors are installed on both sides of the circuit breaker to measure the transient voltage of the circuit breaker. The Rogowski coil sensor is inserted into one end of the monitoring device to collect the transient current flowing through the circuit during the opening and closing process. The signal is collected by the sensor and transmitted to the acquisition card, which is connected to the upper computer software for analysis and then displays the opening and closing time.

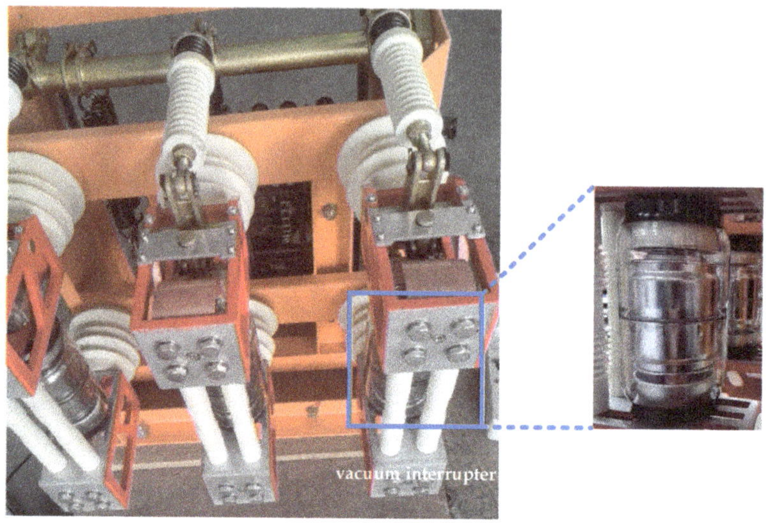

Figure 10. Test the selected circuit breaker.

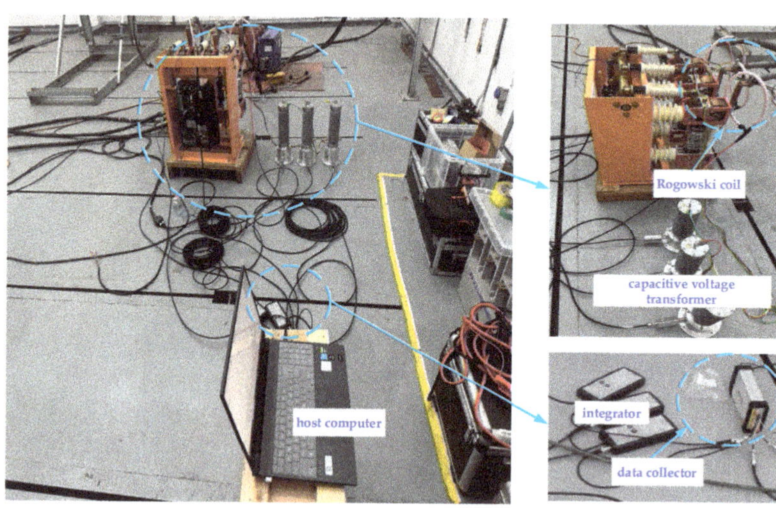

Figure 11. Online monitoring system.

The test conditions in this Section are aimed at measuring transient electrical signals in the closing and opening operating system when the current changes from 13 to 34 A and the power factor angle (cosα) changes from 0.2 to 0.5 when the 400 kVA transformer of the power plant is working at rated state. The main test parameters are shown in Table 2.

Table 2. Test parameter table.

Test Parameter	Opening and Closing Interval Time	Load Parameters
U_N = 10 kV, *In* = 13 A	C-180 ms-O	*R* = 88.83 Ω, *L* = 1385.84 mH cosα = 0.2
U_N = 10 kV, *In* = 20 A		*R* = 86.61, *L* = 877.03 mH cosα = 0.3
U_N = 10 kV, *In* = 27 A		*R* = 85.54 Ω, *L* = 624.16 mH cosα = 0.4
U_N = 10 kV, *In* = 34 A		*R* = 84.91 Ω, *L* = 468.35 mH cosα = 0.5

5.1. Tests Under Different Working Conditions

Figure 12 shows the transient voltage and current waveforms during the opening and closing process of the circuit breaker. The interval between opening and closing is 180 ms, and the circuit breaker is closed first and then opened. From this waveform, it can be observed that the transient process of opening and closing is only a few dozen milliseconds, and the transient voltage and current waveforms collected by the sensor are complete and obvious.

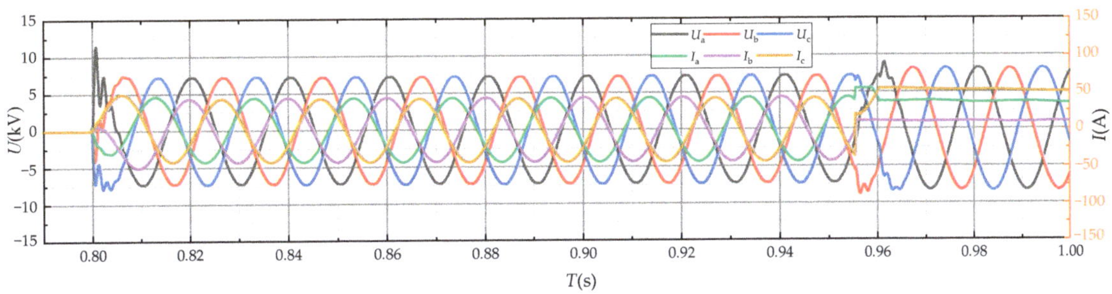

Figure 12. Voltage and current waveforms.

Figure 13 shows the transient voltage and current waveforms in the system during the opening and closing operation when the line voltage is U_N = 10 kV, *In* = 13 A, and cosα = 0.2. Figure 14 shows the voltage and current waveforms in the system during the opening and closing operation when the line voltage is U_N = 10 kV, *In* = 20 A, and cosα = 0.3. During the opening and closing operation, the transient signal characteristics of the system voltage and current are obvious, and it is easy to mark the opening and closing points.

Figure 15 shows the transient voltage and current waveforms in the system during the opening and closing operation when the line voltage U_N = 10 kV, *In* = 27 A, and cosα = 0.4. Figure 16 shows the transient voltage and current waveforms in the system during the opening and closing operation when the line voltage U_N = 10 kV, *In* = 34 A, and cosα = 0.5.

The purpose of collecting the transient electrical signal waveforms of the circuit breaker opening and closing under different working conditions is to prove that different working conditions have transient electrical signal waveforms that are convenient for opening and closing point calibration. The amplitude of the waveform is accidental and will not decrease with the increase in current. For example, the amplitude of Figure 15b is significantly larger than that of Figure 16b. At the same time, the amplitude of the waveform has nothing to do with cosα, but only with the arc generated by the circuit breaker during the opening and closing process.

During the circuit breaker opening and closing test, the transient current may fluctuate to varying degrees due to factors such as the arc cutoff value. However, this does not affect the calibration of the opening and closing time, because the transient waveform selected for the opening and closing time point calibration is the starting and ending points, not the process quantity. At the same time, the transient voltage waveform can be stably measured, and the transient current only serves as an auxiliary judgment.

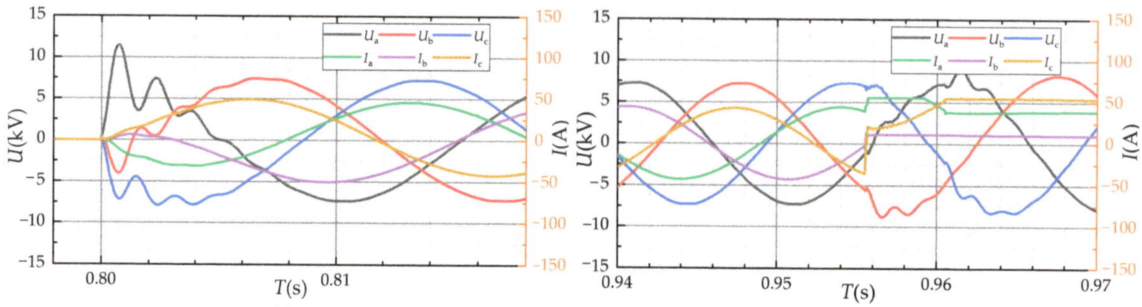

(a) (b)

Figure 13. U_N = 10 kV, *In* = 13 A. Transient voltage and current waveforms in opening and closing test: (**a**) Closing waveform, (**b**) Opening waveform.

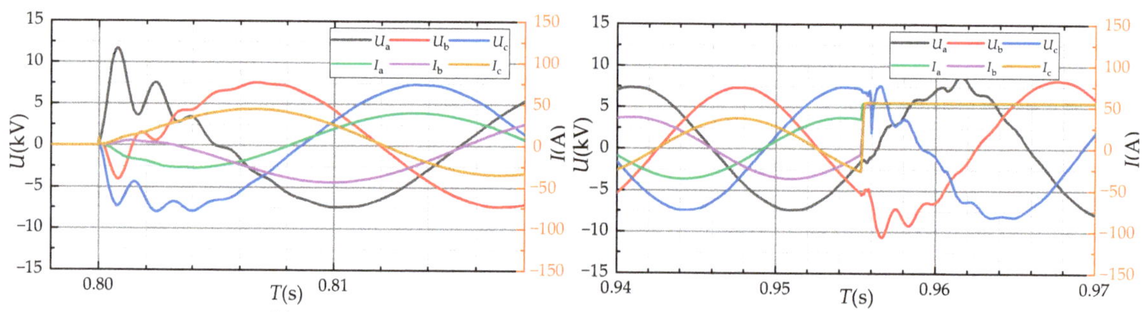

(a) (b)

Figure 14. U_N = 10 kV, *In* = 20 A. Transient voltage and current waveforms in opening and closing test: (**a**) Closing waveform, (**b**) Opening waveform.

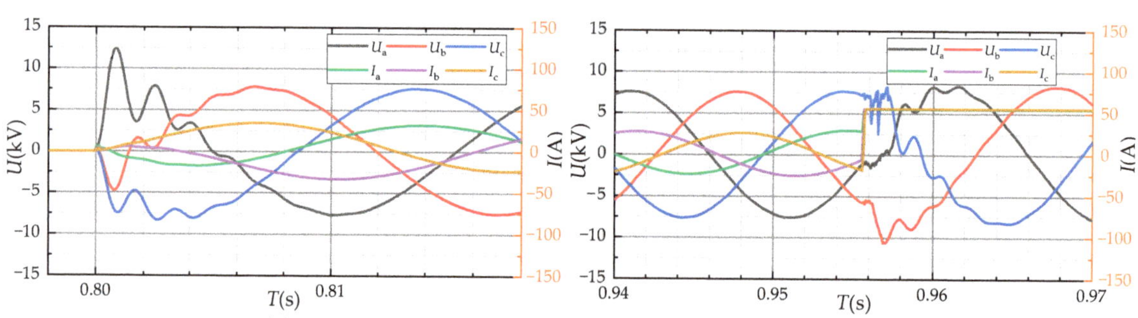

(a) (b)

Figure 15. U_N = 10 kV, *In* = 27 A. Transient voltage and current waveforms in opening and closing test: (**a**) Closing waveform, (**b**) Opening waveform.

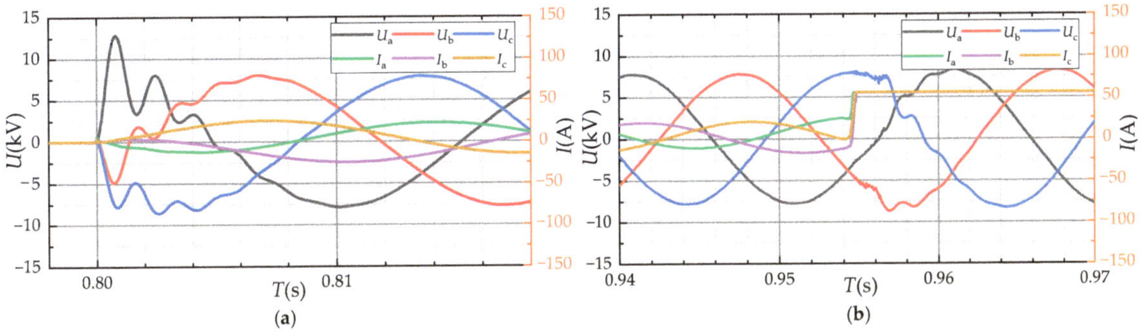

Figure 16. U_N = 10 kV, In = 34 A. Transient voltage and current waveforms in opening and closing test: (**a**) Closing waveform, (**b**) Opening waveform.

The opening and closing time under different working conditions can be calculated by marking the opening and closing start and end points of the opening and closing signals with transient voltage and current waveforms. Transient voltage and current waveforms with obvious opening and closing time feature points are collected under different working conditions. The opening and closing time measurement method based on transient signals is stable, and it can be proved that under different actual working conditions of power plants, the opening and closing time self-calibration can be realized by error compensation, thereby improving the measurement accuracy of the online monitoring system.

5.2. On-Line Measurement of Opening and Closing Time Based on Self-Calibration

In the previous Section, the field test platform was used to prove that the transient voltage and current waveforms can be collected in the circuit breaker opening and closing tests under different working conditions. Phase A of the three-phase circuit breaker with parameters of U_N = 10 kV, In = 27 A, and cosα = 0.4 was selected to carry out a multi-group closing transient electrical signal analysis test.

Before conducting the circuit breaker transient electrical signal test, the circuit breaker mechanical parameter tester was used to measure the closing time–travel curves of multiple groups as shown in Figure 17. The measurement process is to apply a pulse voltage to the opening and closing coils of the circuit breaker through a mechanical stroke tester. The voltage is set to 220 V, and then the displacement–time curve of the moving contact connecting rod is measured by a displacement sensor. The measurement process of each curve is the same, and it is a repeated test under the same parameters. The circuit breaker mechanical parameter tester can stably measure the closing and opening time, and the closing time was measured to be 44.79 ms and the opening time was 29.84 ms, which were used as the benchmark values of the closing and opening time in this system.

Afterward, through 50 groups of closing tests, multiple groups of transient electrical signal waveforms were obtained. Through the opening and closing point calibration method described in the article, the transient waveform was analyzed to determine the just-opening and just-closing points, and finally, the calculated value of the opening and closing time was obtained. The error value of the opening and closing time is based on the difference between the calculated value of the opening and closing points of the transient electrical signal waveform and the reference value measured by the circuit breaker mechanical characteristic tester. Its normal distribution curve is shown in Figure 18.

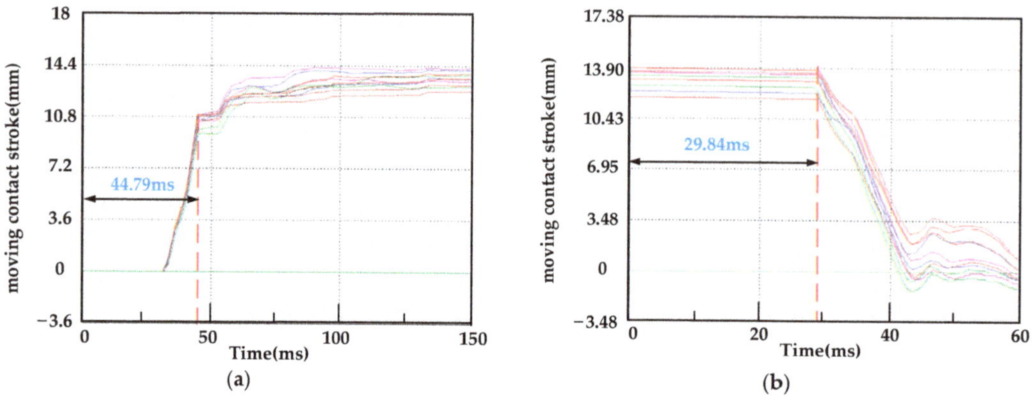

Figure 17. Opening and closing stroke–time curve: (**a**) Closing stroke curve, (**b**) Opening stroke curve.

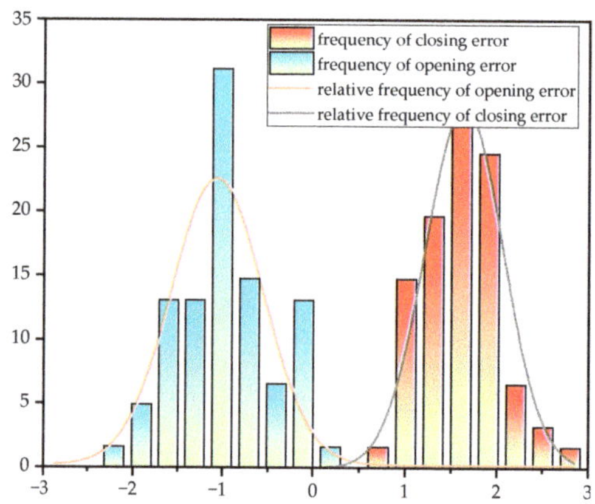

Figure 18. Normal distribution curve of opening and closing time error.

During the opening and closing process, due to certain electromagnetic interference in the environment, there is an impact on the transient electrical signal. The average closing time error is −1.05 ms, and the average opening time error is 1.59 ms.

Then, 30 groups of closing tests were conducted, and the intelligent error compensation module was used for dynamic compensation. The expected value of the closing error was set to 0.5 ms, and the expected value of the opening error was set to 0.5 ms. The LMS algorithm dynamically adjusted the compensation coefficient by minimizing the mean square error between the output signal and the expected signal, so that it can adapt to changes in real time under different environments. The comparison before and after error compensation is shown in Figure 19.

The reference value of the known closing time measured by the mechanical stroke tester is 44.79 ms, and the reference value of the opening time is 29.84 ms. During the 30-group closing test, the average closing time without error compensation is 43.49 ms, its 95% confidence interval is [43.34 ms, 43.66 ms], and the error range is (0.67 ms, 1.77 ms); the average opening time is 30.98 ms, and its 95% confidence interval is [30.83 ms, 31.14 ms], and the error range is (0.49 ms, 2.16 ms). Without error compensation, the goal of measuring accuracy less than 0.5 ms cannot be achieved.

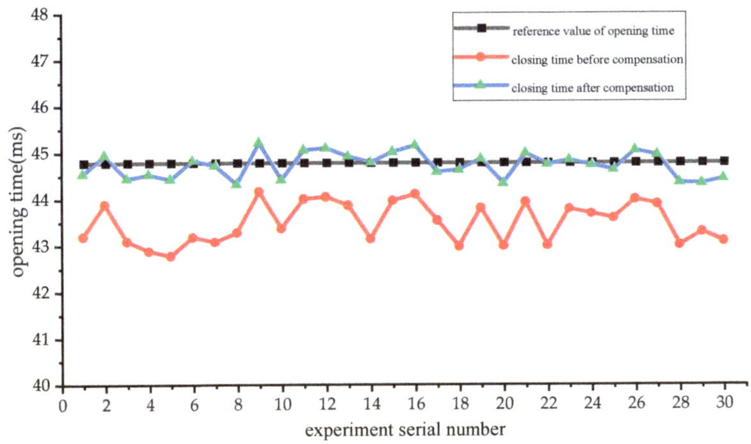

(a)

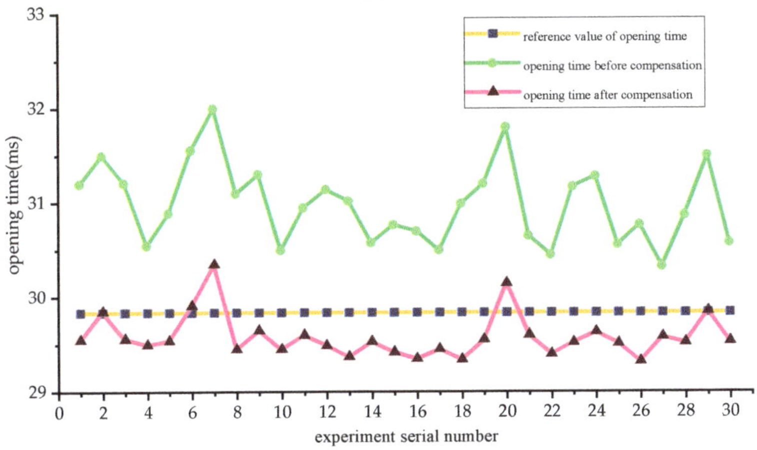

(b)

Figure 19. Opening and closing time error compensation: (**a**) Closing time error compensation, (**b**) Opening time error compensation.

After error compensation, the average closing time is 44.74 ms, its 95% confidence interval is [44.64 ms, 44.84 ms], and the error range is (0.01 ms, 0.48 ms); the average opening time is 29.34 ms, and its 95% confidence interval is [29.20 ms, 29.46 ms], and the error range is (0.01 ms, 0.43 ms). After error compensation, the error between the closing and opening time measured by the online monitoring system based on transient electrical signals and the measurement data of the circuit breaker mechanical characteristics tester is less than 0.5 ms. The closing time measurement results of this method are consistent with the measurement results of the mechanical characteristics tester at 98.93%, and the opening time is consistent with 98.56%. This Section simulates the power plant operating conditions and verifies the stability and accuracy of the method in this paper in measuring the closing and opening time of circuit breakers under power plant operating conditions.

6. Conclusions

This paper innovatively proposes a new online monitoring method for calculating the opening and closing time of spring energy storage vacuum circuit breakers during the opening and closing process, breaking the limitation that the traditional method of mechanical and time characteristic detection must be carried out under power outage and maintenance. This method has high stability, reduces the number of circuit breaker operations required for monitoring, increases the operating life of the circuit breaker, and continuously measures data, which is conducive to fault diagnosis and status monitoring of circuit breakers, providing a guarantee for the safe and reliable operation of the power grid.

First, the transient voltage, current, and the motion trajectory of the moving contact during the closing process of the vacuum circuit breaker are analyzed, and based on the analysis, it is found that the opening and closing information can be extracted from the electrical signal waveform, and then the characteristic quantity (the characteristic points of the "appearance and disappearance" of the transient electrical signal) is extracted, and the self-calibration block is introduced into the opening and closing time calculation program for error compensation. Finally, the opening and closing time samples measured by the experiment are compared with the opening and closing time data obtained by the circuit breaker mechanical characteristics tester to verify the stability and accuracy of the method. The comparison of theoretical research and experimental measurement results shows the following:

- The characteristic quantity extracted by the calculation of this method has high reliability. The online monitoring system measures the opening and closing transient electrical signal waveforms of different working conditions in the power plant environment, and the transient voltage and current waveforms with calibrable characteristic points are collected.
- The opening and closing time calculated by this method, after self-calibration, is consistent with the measurement results of the circuit breaker mechanical characteristics tester by more than 99%.
- The opening and closing time calculation method proposed in this paper is directly related to the transient electrical signal generated by the arc during the opening and closing process of the moving and static contacts of the spring energy storage vacuum circuit breaker. This method can be extended to other switching devices with similar mechanisms.

Author Contributions: Conceptualization, L.W. and P.Z.; methodology, J.X.; software, Y.W.; validation, L.W., Z.Z. and J.W.; formal analysis, L.M.; investigation, X.Z.; resources, L.W.; data curation, J.X.; writing—original draft preparation, L.W.; writing—review and editing, X.Z.; visualization, Z.Z.; supervision, F.Z.; project administration, P.Z.; funding acquisition, L.W. All authors have read and agreed to the published version of the manuscript.

Funding: This research was funded by CHINA YANGTZE POWER Co., Ltd., grant number Z412302043.

Data Availability Statement: The data used in the analysis presented in the paper will be made available, subject to the approval of the data owner.

Conflicts of Interest: Author Yuchuan Wen was employed by the company Three Gorges Ecological Environment Co., Ltd. The remaining authors declare that the research was conducted in the absence of any commercial or financial relationships that could be construed as a potential conflict of interest.

References

1. Janssen, A.; Makareinis, D.; Sölver, C.E. International Surveys on Circuit-Breaker Reliability Data for Substation and System Studies. *IEEE Trans. Power Deliv.* **2014**, *29*, 808–814. [CrossRef]
2. Chen, M.; Lang, B. Study on Opening and Closing Time of Online Monitoring for the Vacuum Circuit Breaker. *High Volt. Appar.* **2015**, *51*, 86–91+97.
3. Strachan, S.M.; McArthur, S.D.J.; Stephen, B.; McDonald, J.R.; Campbell, A. Providing decision support for the condition-based maintenance of circuit breakers through data mining of trip coil current signatures. *IEEE Trans. Power Del.* **2007**, *22*, 178–186. [CrossRef]
4. Razi-Kazemi, A.A.; Vakilian, M.; Niayesh, K.; Lehtonen, M. Circuit-breaker automated failure tracking based on coil current signature. *IEEE Trans. Power Del.* **2014**, *29*, 283–290. [CrossRef]
5. Rudsari, F.N.; Razi-Kazemi, A.A.; Shoorehdeli, M.A. Fault analysis of high-voltage circuit breakers based on coil cur-rent and contact travel waveforms through modified SVM classifier. *IEEE Trans. Power Del.* **2019**, *34*, 1608–1618. [CrossRef]

6. Liu, Q.; Wang, X.; Guo, Z.; Li, J.; Xu, W.; Dai, X.; Liu, C.; Zhao, T. Research on Circuit Breaker Operating Mechanism Fault Diagnosis Method Combining Global-Local Feature Extraction and KELM. *Sensors* **2024**, *24*, 124. [CrossRef]
7. Wang, F.; Gao, W.; Zhao, Y.; Zhang, X.; Yao, S. Condition evaluation of vacuum circuit breaker based on the occurrence of extreme point of vibration signal. *High Volt. Appar.* **2017**, *53*, 248–253.
8. Wu, Y.; Zhang, J.; Yuan, Z.; Chen, H. Fault Diagnosis of Medium Voltage Circuit Breakers Based on Vibration Signal Envelope Analysis. *Sensors* **2023**, *23*, 8331. [CrossRef]
9. Ma, S.; Jia, B.; Wu, J.; Yuan, Y.; Jiang, Y.; Li, W. Multi-Vibration Information Fusion for Detection of HVCB Faults using CART and D–S Evidence Theory. *ISA Trans.* **2021**, *113*, 210–221. [CrossRef]
10. Leone, G.; Cristaldi, L.; Turrin, S. A data-driven prognostic approach based on statistical similarity: An application to industrial circuit breakers. *Measurement* **2017**, *108*, 163–170. [CrossRef]
11. Yang, Q.; Ruan, J.; Zhuang, Z.; Huang, D.; Qiu, Z. Vibration signature extraction of high-voltage circuit breaker by frequency and chaotic analysis. *IEEE Access* **2019**, *7*, 80884–80892. [CrossRef]
12. Yang, Q.; Kong, X.; Wang, Q.; Li, Z.; Cheng, H.; Yu, L. A multi-source ensemble domain adaptation method for rotary machine fault diagnosis. *Measurement* **2021**, *186*, 110213. [CrossRef]
13. Zhou, J.; Luo, Y. Research on Online Monitoring Method of the Instant Closing Point of 12 kV Vacuum Circuit Breaker. *High Volt. Appar.* **2023**, *59*, 219–225.
14. Lu, Y.; Li, Y. Predictive Maintenance Policy for High-voltage Circuit Breakers Based on Real-time State Estimation and Remanent Useful Life Calculation. *High Volt. Eng.* **2022**, *48*, 2716–2726.
15. Guo, N.; Whitmore, K.; Cohen, M.; Beyah, R.; Graber, L. A Non-Invasive Circuit Breaker Arc Duration Measurement Method with Improved Robustness Based on Vibration–Sound Fusion and Convolutional Neural Network. *Energies* **2023**, *16*, 6551. [CrossRef]
16. Da Silva, C.L.; Reis, O.; Assuncao, F.d.O.; Oliveira Castoni, J.C.; Martins, R.; Xavier, C.E.; Areias, I.A.d.-S.; Lambert-Torres, G.; Bonaldi, E.L.; Borges-da-Silva, L.E.; et al. An Online Non-Invasive Condition Assessment Method of Outdoor High-Voltage SF6 Circuit Breaker. *Machines* **2023**, *11*, 323. [CrossRef]
17. Zhang, P.F.; Ding, D.W. Monitoring and analysis of the auxiliary contact bounce of a gas-insulated switchgear cir-cuit breaker according to transient voltage. *J. Tsinghua Univ. (Sci. Technol.)* **2023**, *63*, 1164–1172.
18. Reng, M.; Cheng, R.F. Mechanical and Electrical Fault Diagnosis of Vacuum Circuit Breaker Based on Ternary Plot. *Proc. CSEE* **2022**, *42*, 332–342.
19. Sun, S.G.; Tang, Y. Life Prediction of Operating Mechanism for Circuit Breaker Based on Multiple Signal Feature Fusion. *High Volt. Eng.* **2022**, *48*, 4455–4468.
20. Li, X.; Zhu, K.; Guo, Z.; Zhang, Y.; Jia, S.; Jiang, X. Experimental study on arc interruption characteristics of SF6 and its mixtures with CF4. *Proc. CSEE* **2017**, *37*, 3315–3322.
21. Razi-Kazemi, A.A.; Niayesh, K.; Nilchi, R. A Probabilistic Model-Aided Failure Prediction Approach for Spring-Type Operating Mechanism of High-Voltage Circuit Breakers. *IEEE Trans. Power Deliv.* **2019**, *34*, 1280–1290. [CrossRef]

energies

MDPI

Article

Online Measurement Method for Circuit Breaker Mechanical–Time Characteristics Based on Transient Voltage and Current Signal Feature Extraction

Liting Weng [1], Jianhua Xia [1], Zhaochuang Zhang [1], Jingang Wang [2], Lin Chen [3], Yingbo Zi [3], Lingyi Ma [2], Xingyu Zhang [2], Fan Zhang [2] and Pengcheng Zhao [2,*]

[1] Xiluodu Hydropower Plant, Zhaotong 657300, China; weng_liting@ctg.com.cn (L.W.); xia_jianhua@ctg.com.cn (J.X.); zhang_zhaochuang@ctg.com.cn (Z.Z.)
[2] State Key Laboratory of Power Transmission Equipment Technology, School of Electrical Engineering, Chongqing University, Chongqing 400044, China; jingang@cqu.edu.cn (J.W.); 202311131275@stu.cqu.edu.cn (L.M.); 202311131307@stu.cqu.edu.cn (X.Z.); zhangfan023@stu.cqu.edu.cn (F.Z.)
[3] Three Gorges Ecological Environment Co., Ltd., Zhaotong 657300, China; chen_lin5@ctg.com.cn (L.C.); zi_yingbo@ctg.com.cn (Y.Z.)
* Correspondence: zhaopengcheng@cqu.edu.cn

Abstract: The good mechanical characteristics of circuit breakers are essential to ensuring their safe and stable operation. The characteristic signals at the points of initial opening and closing contain abundant information about the mechanical–time characteristics of the circuit breaker. Existing online measurement methods for circuit breaker mechanical–time characteristics typically rely on monitoring the secondary current signals of the opening and closing coils, or the signals such as vibration, pressure, and angular displacement that are generated during the opening and closing of moving and stationary contacts. Unlike these conventional approaches, this paper proposes an online measurement method for circuit breaker mechanical–time characteristics based on feature extraction from transient voltage and current signals in the primary circuit. By extracting features from the transient voltage and current signals in the primary circuit, this method obtains mechanical–time characteristics such as the points of initial opening and closing. This method is convenient and stable, enabling the online detection of the mechanical characteristics of the circuit breaker. Finally, a comparison and analysis of the online measurement results with the offline measurement results were conducted. The analysis shows that the calculation error of the opening and closing times is within 0.2 ms. This online detection method has a high measurement accuracy and provides a new approach for the online measurement of the mechanical characteristics of circuit breakers.

Keywords: circuit breaker; mechanical–time characteristics; online measurement; transient voltage; current signals

1. Introduction

High-voltage circuit breakers are switching devices that play a critical role in controlling and protecting the power grid during operation. The accuracy and reliability of their operation are closely tied to the safe operation of the entire system. However, current statistics indicate that more than half of major incidents are caused by the mechanical failure of circuit breakers. Therefore, it is particularly important to strengthen the online detection of circuit breaker mechanical–time characteristics [1,2]. The online detection of circuit breaker mechanical–time characteristics involves collecting operational data from the circuit breaker, extracting characteristic parameters, and establishing diagnostic models to achieve an online evaluation of the circuit breaker's operational status.

Currently, the characteristic signals studied in circuit breaker mechanical–time characteristics mainly include opening and closing coil currents, vibration signals, pressure signals, and contact displacement signals. By calculating the first derivative of the closing coil current, the initial closing point of the circuit breaker can be determined, but this method has limited measurement accuracy [3]. The authors of [4] proposed a method for detecting the opening and closing coil currents of circuit breakers using a Hall-effect CT current clamp and three clamp-on current sensors. This approach enables the qualitative detection of a circuit breaker's mechanical characteristics, improving measurement accuracy; however, it introduces errors from multiple sensors. When using vibration signals to detect the mechanical state of circuit breakers, the short-time energy method is typically employed for the time domain analysis of the vibration signals, which can effectively extract the initial opening and closing points of the circuit breaker. However, this method introduces measurement errors from the vibration sensors [5,6]. By combining the acquisition and analysis of vibration signals and opening/closing coil current signals, the online monitoring of mechanical faults in a circuit breaker's operating mechanism can be achieved, thereby improving measurement accuracy [7,8]. When using pressure sensors for the online detection of a circuit breaker's mechanical–time characteristics, the first extremum point after the pressure surge is selected as the initial closing point, but this method also introduces sensor errors [9,10]. In methods that use angular displacement sensors to detect a circuit breaker's mechanical characteristics, the relationship between the spindle rotation angle and the linear displacement of the contacts is established based on an analysis of the circuit breaker's linkage structure. Experimental validation is then performed to verify the accuracy of using angular displacement sensors for contact monitoring [11,12]. The method for extracting the initial closing point based on the circuit breaker's travel–time curve uses the closing speed curve as key information. Wavelet decomposition and reconstruction are then applied to the closing speed curve to extract characteristic parameters and calculate the initial closing point [13].

This paper conducts an online detection of a circuit breaker's mechanical–time characteristics based on the transient voltage and current signals generated in the primary circuit during circuit breaker opening and closing operations. These signals are easy to collect, exhibit distinct characteristics, and contain abundant information about the mechanical–time characteristics of the circuit breaker, directly reflecting its actual operating condition during fault current or load current switching. This paper establishes the relationship between transient electrical signals and circuit breaker mechanical–time characteristics through the collection and analysis of transient voltage and current signals in the primary circuit during switching operations. By using feature extraction algorithms, time characteristics such as the initial opening and closing points of the circuit breaker are obtained. Finally, the accuracy of this online detection method is verified through experiments. Compared to traditional offline detection methods, the relative error in extracting the initial opening and closing points is within 0.2 ms. This provides a new research approach for the online detection of a circuit breaker's mechanical–time characteristics, offering a new direction for future research in a circuit breaker's online fault diagnostics.

2. Transient Characteristic Analysis of Circuit Breaker Switching Electrical Signals

2.1. Theoretical Analysis of Transient Electrical Signals During Circuit Breaker Switching

Vacuum circuit breakers are widely used in medium-voltage systems and offer a series of advantages such as fast switching, small size, and a light weight. However, practical applications show that vacuum circuit breakers are prone to arc breakdown and severe switching overvoltage during operation. The mechanical characteristics of the circuit breaker, contact materials, and other factors are closely related to this overvoltage phenomenon [14]. A typical view of a vacuum interrupter chamber is shown in Figure 1.

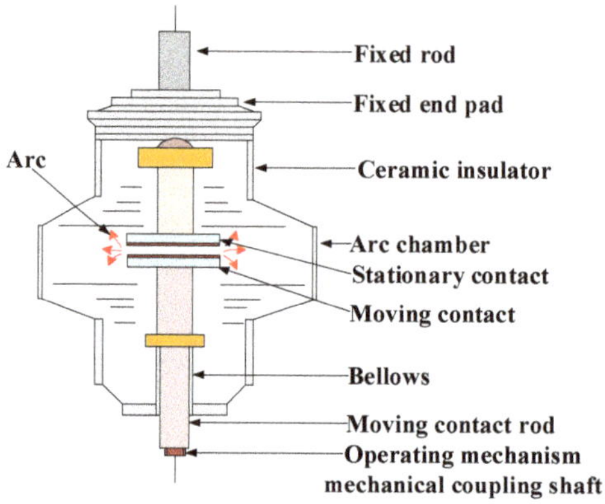

Figure 1. Typical perspective view of a vacuum arc extinguishing chamber.

The following parameters should be emphasized during the circuit breaker's opening and closing processes:

(1) Chopping value: the circuit breaker's ability to naturally interrupt the current and pass through zero.
(2) Dynamic dielectric strength: the characteristic of the insulation strength recovery between contacts during the circuit breaker's operation.
(3) Transient recovery voltage (TRV): the significant recovery voltage that appears across the contacts after the arc extinction.
(4) High-frequency arc extinguishing ability: the circuit breaker's ability to extinguish high-frequency currents as they pass through zero.

Currently, the average value method is primarily used to characterize the natural current–zero capability of circuit breakers. The average chopping current is defined as follows:

$$I_{\mathrm{ch}} = (\omega I_0 \alpha \beta)^q \tag{1}$$

where ω represents the angular frequency; I_0 is the amplitude of the power frequency current; and α, β, and q are the contact material parameters. For Cu/Cr contacts, the parameters are $\alpha = 6.2 \times 10^{-16}$ s, $\beta = 14.2$, and $q = (1 - \beta) - 1$.

During the circuit breaker's opening operation, the dynamic dielectric strength of the medium between the contacts is considered to have a linear relationship with time, as follows:

$$U_{\mathrm{b0}} = A(t - t_0) + B \tag{2}$$

where U_{b0} represents the average withstand voltage of the vacuum circuit breaker, t_0 is the moment of contact separation, A denotes the rate of increase in the dielectric strength of the medium in the vacuum circuit breaker, and B represents the transient recovery voltage before the current zero crossing [15].

The TRV is significantly influenced by the arc, and the nonlinear arc Schwarz–Avdonin model is based on Equations (3)–(5):

$$\frac{dg}{dt} = \frac{1}{\tau(g)}[\frac{i^2}{P(g)} - g] \tag{3}$$

$$P = P_0 g^\beta \tag{4}$$

$$\tau = \tau_0 g^\alpha \tag{5}$$

where g is the arc conductance (S/m); i is the arc current (A); τ is the thermal time constant of the arc conductance (s); P is the cooling power of the arc conductance (W); and α and β are free parameters [16].

The high-frequency arc-extinguishing capability of the circuit breaker is closely related to whether the arc can be extinguished during interruption, and is expressed as follows:

$$dI/dt = C(t - t_0) + D \tag{6}$$

where C represents the rise rate characterizing the high-frequency arc-extinguishing capability, and D represents the arc-extinguishing capability.

The dynamic insulation strength during pre-breakdown caused by circuit breaker closing is modified as follows:

$$U_{b0} = 1.4 \times 1.5 \times \sqrt{2/3} U_{rated} - [A(t - t_1) + B] \tag{7}$$

where U_{rated} is the rated voltage of the circuit breaker, and t_1 is the closing moment of the circuit breaker [17].

From the above theoretical analysis, it can be concluded that high-frequency voltage and current transient signals are generated during the operation of the circuit breaker. The following section will further analyze the transient electrical signals generated during the circuit breaker's operation through simulation and will perform a time–frequency analysis.

2.2. Simulation Analysis of Transient Electrical Signals During Circuit Breaker Switching Operations

To further perform a time–frequency analysis of the re-ignition during breaker opening and the pre-breakdown waveform during closing, this paper constructs a "vacuum circuit breaker–cable–load" system simulation model, as shown in Figure 2, within the EMTP-ATP simulation environment.

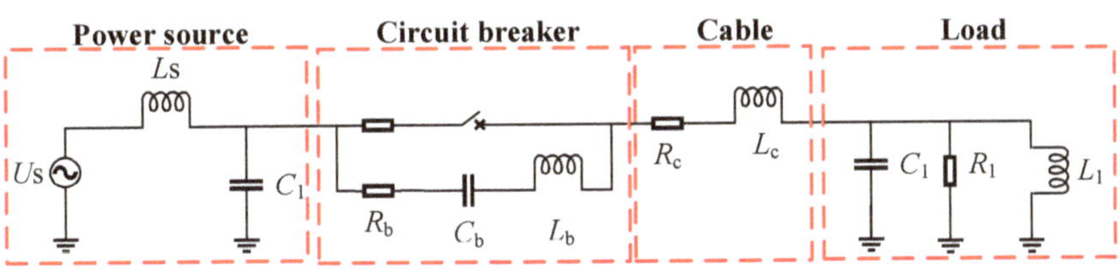

Figure 2. "Vacuum circuit breaker–cable–load" system simulation model.

The simulation model includes modules for the power source, circuit breaker, cable, and load, with simulation parameters shown in Table 1. Among these, U_s is the effective value of the power supply voltage, and L_s and C_s represent the source reactance, characterizing the capacitive and inductive effects of the power source. The values of L_s and C_s are set to 0.1 mH and 0.1 μF, respectively, which fall within a reasonable range for medium-voltage distribution networks and are easy to match with other system parameters. R_b, C_b, and L_b represent the circuit breaker impedance, characterizing the stray resistance, capacitance and inductance between the contacts. The stray resistance R_b is set to 50 Ω, which reasonably reflects the contact resistance and current path loss between the contacts. C_b and L_b represent the electric and magnetic fields between the breaker contacts. Due to the small contact area, these values are typically small. R_c and L_c represent the cable impedance, with R_c set to 2 Ω and L_c set to 0.04 mH, which reflect the cable length from the distribution room to the load in a medium-voltage distribution system. C_1, R_1, and L_1 represent the load impedance [2,18].

Table 1. Simulation parameters.

Parameter	Value
U_s	8.16 kV
L_s	0.1 mH
C_s	0.1 μF
R_b	50 Ω
C_b	0.2 nF
L_b	0.05 nH
R_c	2 Ω
L_c	0.04 mH
C_1	0.01 μF
R_1	10,000 Ω
L_1	120 mH

The input parameters of the VBM code are source voltage u_2, load voltage u_2, and circuit breaker current i_k. The output parameter is the breaking signal trip. During each simulation time step, the specific calculation process is as follows:

(1) Set the computation time of EMTP-ATP as t. At this moment, the source voltage is u_1, the load voltage is u_2, and the circuit breaker current is i_k.
(2) Compare t with the preset separation time t_0 of the vacuum circuit breaker. If the condition is met, calculate ich, U_b, and di/dt; otherwise, skip the calculation and move to the next time step.
(3) First, determine whether chopping occurs in the first interrupting phase. Then, sequentially compare the current value *i(t)* at time t with the chopping current ich, the current zero-crossing slope *K*, di/dt, and the relationship between $|u_1-u_2|$ and U_b. Finally, output the breaking signal trip.

For pre-breakdown during closing operations, only the parameters of insulation strength and arc-extinguishing capability need to be considered.

In the simulation model shown above, the circuit breaker opening and closing operations are executed, yielding the transient voltage and current waveforms, as illustrated in Figures 3 and 4.

In the simulation results shown in Figure 3, the circuit breaker opening operation command is issued at 7 ms. At 7.43 ms (point B), the current flowing through the circuit breaker drops below the chopping current I_{ch}, causing a power frequency current interruption and triggering a rapid rise in the transient recovery voltage (TRV). Between 7.548 ms (point C) and 7.551 ms (point D), the TRV repeatedly exceeds the dynamic dielectric strength U_{b0}, leading to arc reignitions, while the high-frequency arc current undergoes multiple interruptions, resulting in a fluctuating TRV. During this period, the maximum transient voltage reaches 34.086 kV with a frequency of 1.667 MHz, and the transient current frequency is 3.33 MHz. At point D, a high-frequency current interruption occurs, causing the TRV to rise sharply before gradually decreasing, with the current being temporarily interrupted [19,20]. After repeating this process four times, the arc current is successfully interrupted at point E, and the circuit breaker achieves a successful opening operation.

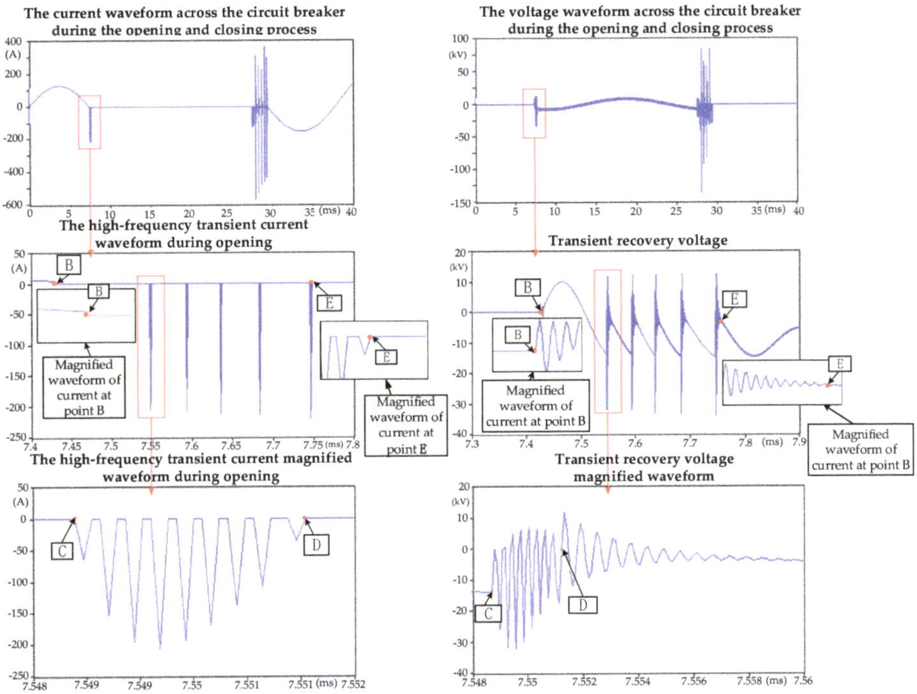

Figure 3. Analysis of opening waveforms in the simulation of the "vacuum circuit breaker–cable–load" system.

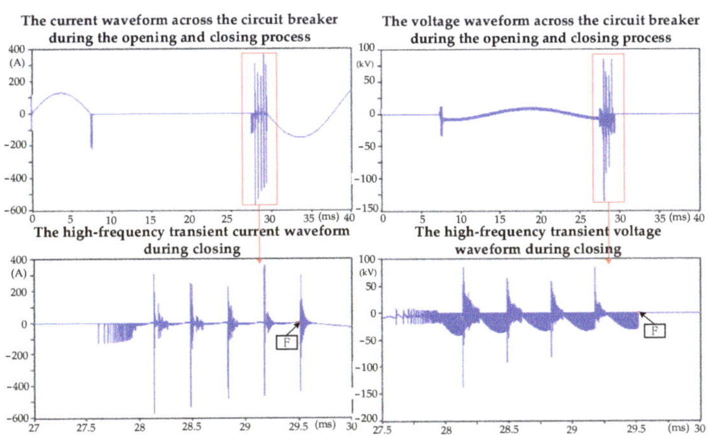

Figure 4. Analysis of closing waveforms in the simulation of the "vacuum circuit breaker–cable–load" system.

In the simulation results shown in Figure 4, a circuit breaker closing operation command is issued at 27 ms. At this time, the dynamic insulation strength U_{b0} between the circuit breaker contacts decreases over time, leading to multiple pre-breakdown phenomena. This process is accompanied by high-frequency currents similar to those observed during restrikes. The maximum transient closing voltage reaches 136.8 kV with a frequency of 5 MHz, while the transient closing current frequency is 0.25 MHz. At 29.525 ms (point F), the moving and stationary contacts of the circuit breaker make full contact, completing the closing operation. The voltage drops sharply to zero, causing high-frequency oscillations in the current, which subsequently attenuate to the power frequency.

2.3. Characteristic Extraction Method for the Circuit Breaker's Initial Opening and Closing Points

Based on the above theoretical and simulation analysis of the transient signals during circuit breaker operation, this paper proposes a method for extracting the features of the circuit breaker's initial opening and closing points using transient voltage and current signals. The detailed process is shown in Figure 5.

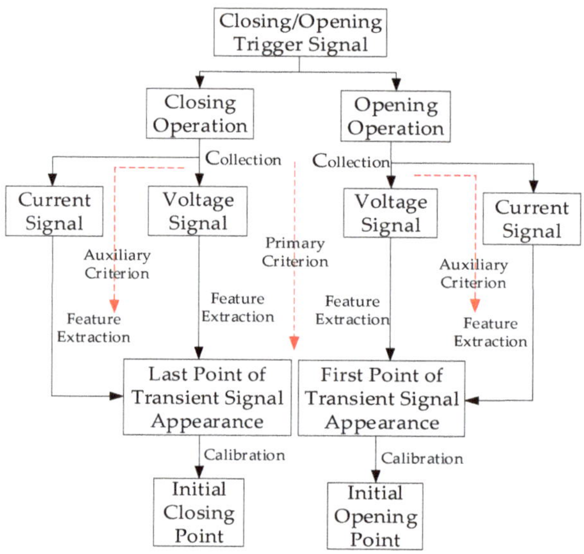

Figure 5. Method for extracting initial opening and closing points.

When the closing/opening signal is received, the circuit breaker executes the corresponding closing/opening operation. The PT and CT in the system collect the voltage and current signals at both ends of the circuit breaker. Meanwhile, the feature extraction algorithm processes the collected signals. During closing operations, the last detected point of the transient voltage/current signal is identified as the initial closing point. During opening operations, the first detected point of the transient voltage/current signal is identified as the initial opening point. The voltage signal collected by the PT serves as the primary criterion, while the current signal collected by the CT serves as the auxiliary criterion.

First, wavelet denoising is used to remove system noise from the voltage and current waveforms collected by the voltage and current sensors. Then, slope scanning is performed on the collected waveforms. Since the frequency of transient electrical signals is significantly higher than that of power frequency signals, the slope scanning algorithm can accurately identify the points of initial separation and closure.

Figures 6 and 7 illustrate the calibration of the opening and closing points, using the voltage as the primary criterion and the current as the auxiliary criterion.

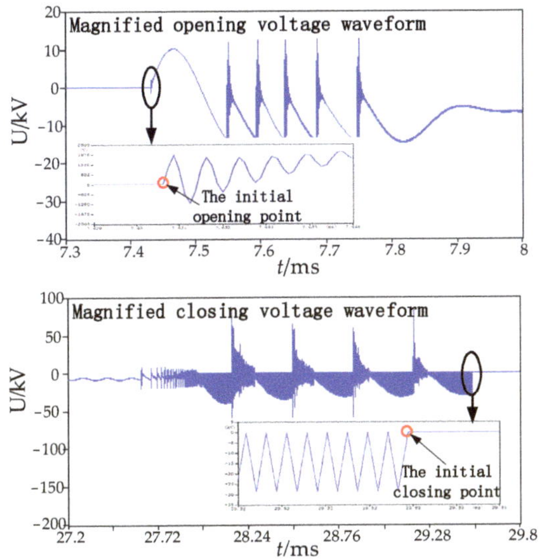

Figure 6. Opening and closing point calibration based on the voltage as the primary criterion.

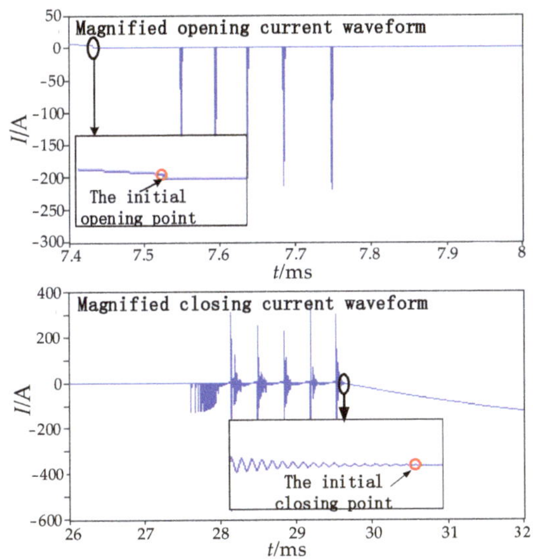

Figure 7. Opening and closing point calibration based on the current as the auxiliary criterion.

The extraction results of the opening and closing points based on different voltage and current criteria are very similar, but the opening and closing points based on the transient voltage signal are easier to extract (Table 2). Furthermore, in the subsequent circuit breaker opening and closing tests, it was found that when a vacuum circuit breaker interrupts a small current inductive load, the transient current signal is not obvious, while the transient voltage signal is more pronounced. Therefore, the transient voltage signal is selected as the primary criterion, and the transient current signal is used as the auxiliary criterion.

Table 2. The extraction results of the opening and closing points based on different voltage and current criteria.

	The Initial Opening Point	The Initial Closing Point
Voltage signal	0.0074308 ms	0.029525 ms
Current signal	0.0074306 ms	0.029632 ms

3. Test and Validation

3.1. Offline Measurement of Opening and Closing Times

In this study, the breaker's initial opening and closing points, measured by a mechanical characteristic tester, are used as a reference to calculate the relative error of the proposed online measurement method.

The mechanical characteristic tester can synchronize the acquisition of the breaker's contact electrical signal and the stroke–time curve during the measurement of the breaker's opening and closing stroke–time curves. By amplifying the contact electrical signal, its intersection point with the stroke–time curve is determined, representing the breaker's initial opening and initial closing points [21]. Figures 8 and 9 show the offline measurement results of the breaker's initial opening and initial closing points, which were obtained using the mechanical characteristic tester.

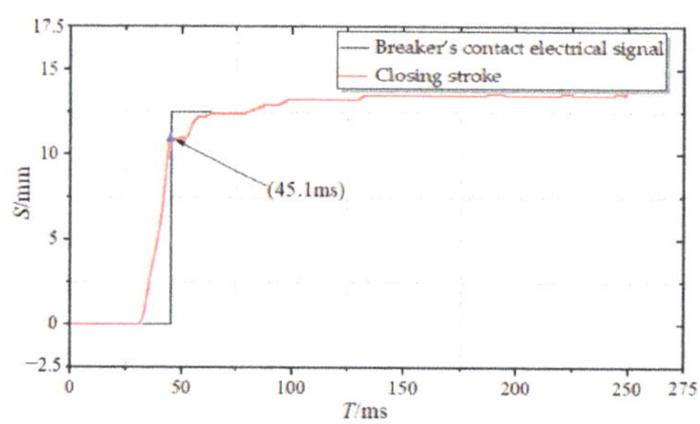

Figure 8. Offline measurement results of the closing point.

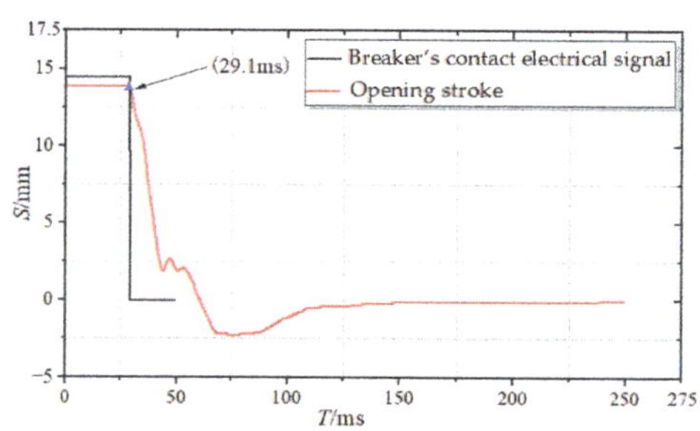

Figure 9. Offline measurement results of the opening point.

A total of 50 offline opening and closing tests were conducted on the circuit breaker using a mechanical characteristic tester. The average closing time measured from the 50 tests was 45.21 ms, and the average opening time was 29.43 ms. These values will be used as the reference to calculate the relative error of the online measurement method in the following discussion.

3.2. Online Acquisition System for Circuit Breaker Mechanical–Time Characteristics

To verify the accuracy of the online measurement method proposed in this paper, an online testing platform was set up as shown in Figure 10.

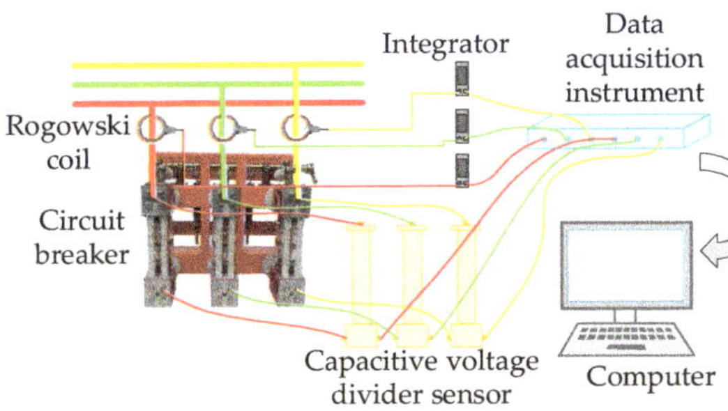

Figure 10. Structural Diagram of the Circuit Breaker Mechanical-Time Characteristics Online Testing Platform.

Figure 11 is the on-site layout of the circuit breaker mechanical–time characteristic online testing simulation platform. The test control console, with an output voltage range of 0–250 V, is connected to a step-up transformer to produce an output voltage of 0.2–50 kV. The output terminals of the Rogowski coil and the capacitive voltage divider are connected to the data acquisition instrument, which, in turn, is linked to the industrial control computer to display the voltage and current waveforms. The Rogowski coil has a response bandwidth of 30 Hz–1 MHz, the passive integrator has a response bandwidth of 20 Hz–2 MHz, and the capacitive voltage divider has a response bandwidth of 30 Hz–10 MHz, meeting the measurement requirements for the field test.

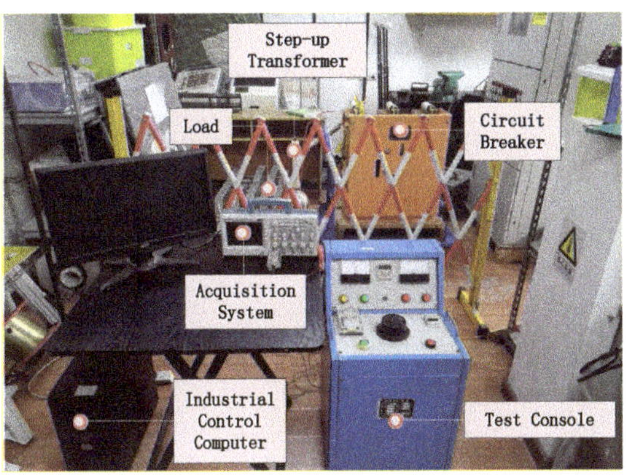

Figure 11. The on-site layout of the circuit breaker mechanical–time characteristic online testing simulation platform.

3.3. Repeatability Testing and Results Analysis

In this paper, a mechanical characteristic tester is used to provide the tripping trigger pulse [22], and 60 rounds of circuit breaker tripping repeatability tests are carried out on the constructed test platform; the test conditions are U_n = 10 kV, I_n = 50 A, and cosα = 0.8 (hysteresis). Among them, the closing pulse signal is connected to channel 1 of the data acquisition instrument, the breaking signal is connected to channel 2, the voltage signal is connected to channel 3, and the current signal is connected to channel 4. Figure 12 shows the circuit breaker opening and closing waveforms obtained from any one online testing platform.

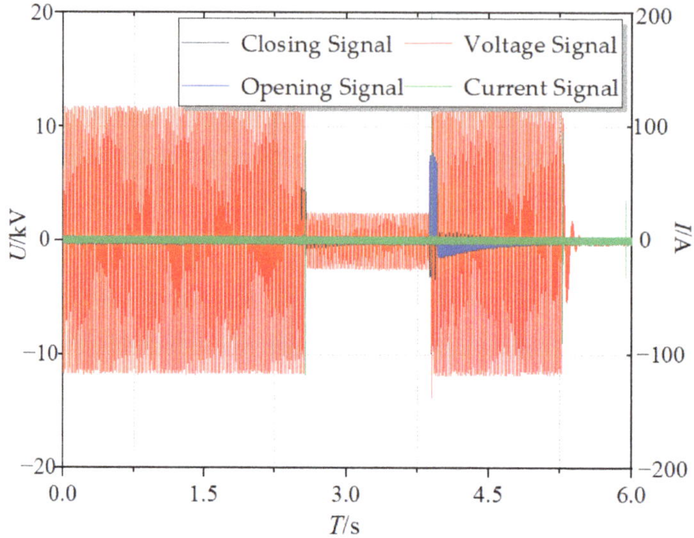

Figure 12. The opening and closing waveform of the circuit breaker measured by the online testing platform.

Feature extraction was performed on the 60 sets of breaker switching waveforms obtained, and the initial opening and closing points, as well as the opening and closing times, for each test were determined. The calculation method for the opening and closing times is shown in Figures 13 and 14.

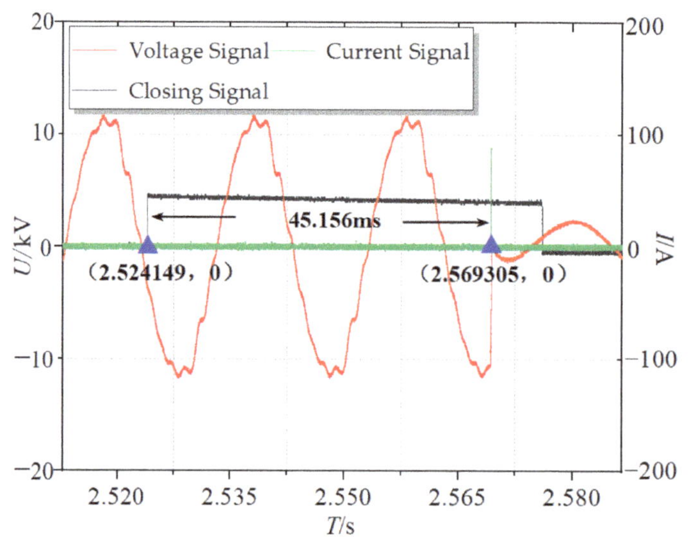

Figure 13. Calculation method for circuit breaker closing time.

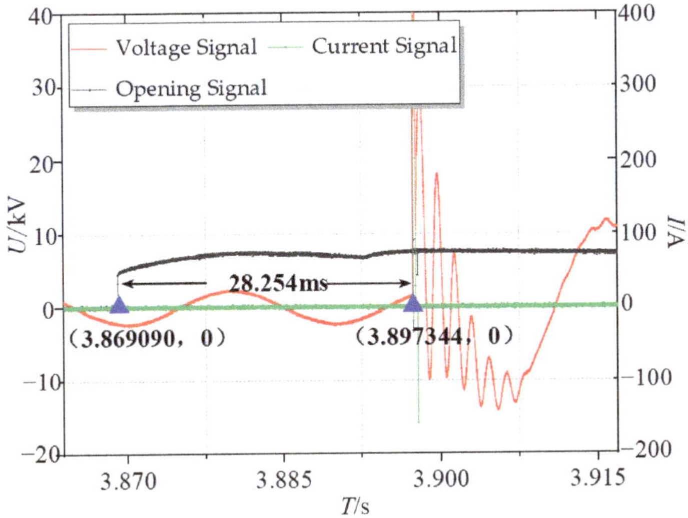

Figure 14. Calculation method for circuit breaker opening time.

As shown in Figures 13 and 14, the breaker closing time is the difference between the last appearance time of the closing transient voltage/current signal and the closing signal trigger time; the breaker opening time is the difference between the first appearance time of the opening transient voltage/current signal and the opening signal trigger time.

The 60 sets of closing and opening time data obtained based on the above method are compared with the closing time of 44.79 ms and opening time of 29.84 ms to calculate the relative error of the online measurement of the closing and opening times. The errors of any 10 sets of closing and opening times are shown in Tables 3 and 4. The closing and opening times from the 60 sets of repeatability tests are plotted as normal distribution charts in Figures 15 and 16; error analysis is conducted on the test data.

Table 3. Online test closing time error statistics table.

Serial Number	Closing Trigger Time (ms)	Feature Extraction Closing Point (ms)	Online Test Closing Time (ms)	Closing Time Error (ms)
1	969.841	1014.949	45.108	0.118
2	2187.064	2232.209	45.145	0.155
3	1877.613	1922.810	45.197	0.107
4	2524.140	2569.302	45.162	0.172
5	2583.509	2628.798	45.289	0.199
6	2046.359	2091.510	45.151	0.161
7	1310.238	1355.399	45.161	0.171
8	2719.299	2764.633	45.334	0.144
9	944.216	989.246	45.03	0.24
10	1374.799	1420.009	45.21	0.12

Table 4. Online test opening time error statistics table.

Serial Number	Opening Trigger Time (ms)	Feature Extraction Opening Point (ms)	Online Test Opening Time (ms)	Opening Time Error (ms)
1	4921.534	4950.653	29.119	−0.121
2	5603.575	5633.297	29.722	−0.118
3	2191.749	2221.295	29.546	−0.294
4	4123.226	4153.035	29.809	−0.031
5	2235.563	2264.819	29.256	−0.184
6	2450.111	2480.035	29.924	0.084
7	6330.277	6360.003	29.726	−0.114
8	3536.032	3566.013	29.981	0.141
9	3841.710	3871.508	29.798	−0.042
10	5178.455	5207.043	28.588	−0.252

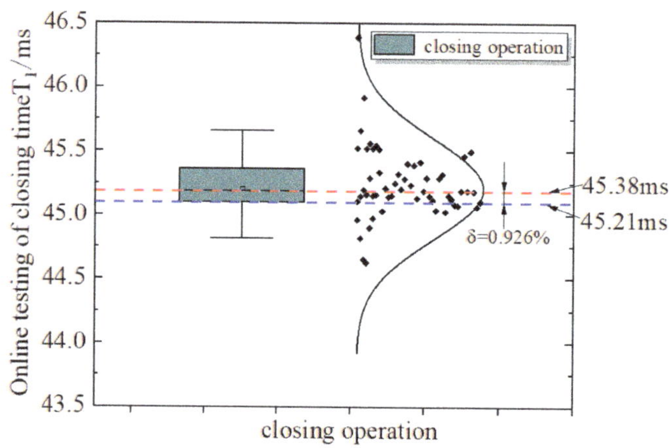

Figure 15. Normal distribution diagram of closing time in 60 repeated tests.

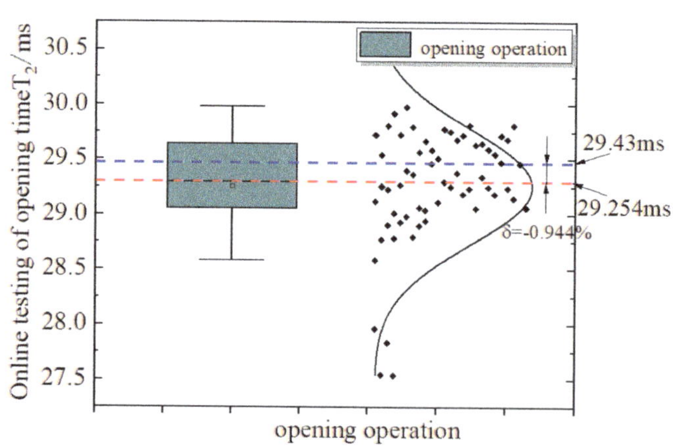

Figure 16. Normal distribution diagram of opening time in 60 repeated tests.

The formulas for calculating the mean (ME), variance (s^2), mean absolute deviation (MAD), and relative error (δ) of the breaker opening and closing times obtained from 60 sets of online tests are as follows:

$$\begin{cases} \mathrm{ME}_{\mathrm{close}} = \dfrac{\sum\limits_{n=1}^{60} T_{1n}}{60} \\ \mathrm{ME}_{\mathrm{open}} = \dfrac{\sum\limits_{n=1}^{60} T_{2n}}{60} \end{cases} \tag{8}$$

$$\begin{cases} s_{\mathrm{close}}^2 = \dfrac{1}{60}\sum\limits_{n=1}^{60} (T_{1\mathrm{n}} - \mu_1)^2 \\ s_{\mathrm{open}}^2 = \dfrac{1}{60}\sum\limits_{n=1}^{60} (T_{2\mathrm{n}} - \mu_2)^2 \end{cases} \tag{9}$$

$$\begin{cases} \mathrm{MAD}_{\mathrm{close}} = \dfrac{\sum\limits_{n=1}^{60} |T_{1\mathrm{n}} - \mu_1|}{60} \\ \mathrm{MAD}_{\mathrm{open}} = \dfrac{\sum\limits_{n=1}^{60} |T_{2\mathrm{n}} - \mu_2|}{60} \end{cases} \tag{10}$$

$$\begin{cases} \delta_{\mathrm{close}} = \frac{\mathrm{ME}_{\mathrm{close}} - \mu_1}{\mu_1} \times 100\% \\ \delta_{\mathrm{open}} = \frac{\mathrm{ME}_{\mathrm{open}} - \mu_2}{\mu_2} \times 100\% \end{cases} \tag{11}$$

where, μ_1= 45.21 ms and μ_2= 29.43 ms. The calculation results are shown in Table 5.

Table 5. Analysis table of online test results for opening and closing times.

	ME (ms)	s^2 (ms)	MAD (ms)	δ (ms)
Opening time	45.38	0.126	0.185	0.926%
Closing time	29.245	0.182	0.139	−0.944%

From the analysis of the data in the table above, it can be seen that although there is some fluctuation in the 60 sets of data, the overall values are controlled within a certain range. The variance of the closing time is 0.126 ms, and the variance of the opening time is 0.182 ms, both within 0.2 ms, indicating small data fluctuations. The mean absolute deviation of the closing time is 0.185 ms, and the mean absolute deviation of the opening time is 0.139 ms, showing low data dispersion. The relative error of the closing time for the 60 sets of data is 0.926%, and the relative error of the opening time is −0.944%, both controlled within ±1%, indicating that the measurement method has high accuracy. Based on the error analysis of the 60 sets of online test results, it can be concluded that the online measurement method proposed in this paper offers a high measurement accuracy and good stability.

4. Conclusions

This paper proposes an online measurement method for the breaker's initial opening and closing points based on the extraction of transient voltage and current signal characteristics. Firstly, the time–frequency characteristics of the transient electrical signals during the breaker's opening and closing processes are studied through theoretical analysis and simulation modeling. Secondly, a breaker's opening and closing test platform is set up to conduct multiple sets of online measurement experiments. The proposed initial opening and closing point online measurement method is then used to extract features from the measurement results, which are compared with the offline measurement results. The main conclusions of this paper are as follows:

(1) The variance and mean absolute deviation of the opening/closing times measured by the proposed instantaneous opening and closing point online measurement method in this paper are both within 0.2 ms, and the relative error of the opening or closing time is within ±1%. The measurement results show low fluctuation and high stability.
(2) This online measurement method obtains the time characteristic information of the breaker without affecting the normal operation of the system. Compared to traditional offline measurement methods using mechanical characteristic testers, this online measurement method offers stronger timeliness and continuity.
(3) This online detection system is based on a completely new approach, expanding the research field of online measurements of breaker mechanical and time characteristics. Compared to other online detection methods, this online measurement method innovatively extracts the characteristic information of the instantaneous opening and closing points of the breaker by analyzing the transient electrical signals generated during the breaker's opening and closing processes in the primary circuit. This provides a new research direction for the subsequent online detection of breaker time characteristics.

Author Contributions: Conceptualization: L.W. and P.Z.; methodology: J.X.; software: Z.Z.; validation: J.W., L.C. and Y.Z.; formal analysis: L.M.; investigation: X.Z.; resources: F.Z.; data curation: L.W.; writing—original draft preparation: L.M.; writing—review and editing: X.Z.; visualization: Z.Z.; supervision: J.W.; project administration: P.Z.; funding acquisition: P.Z. All authors have read and agreed to the published version of the manuscript.

Funding: This research was funded by China Yangtze Power Co., Ltd.; grant number Z412302043.

Data Availability Statement: The data used in the analysis presented in the paper will be made available, subject to the approval of the data owner.

Conflicts of Interest: Authors Lin Chen and Yingbo Zi were employed by the company Three Gorges Ecological Environment Co., Ltd. The remaining authors declare that the research was conducted in the absence of any commercial or financial relationships that could be construed as a potential conflict of interest.

References

1. Yang, F.; Wang, X.; Rong, M. A novel on-line monitoring method for three phase synchronization of medium voltage vacuum circuit breaker. *Proc. Chin. Soc. Electr. Eng.* **2008**, *28*, 139.
2. Geng, Y.; Yao, X.; Dong, J.; Liu, X.; Geng, Y.; Liu, Z.; Peng, J.; Wang, K. Experimental Investigation of the Prestrike Characteristics of a Double-Break Vacuum Circuit Breaker under DC Voltages. *Energies* **2020**, *13*, 3217. [CrossRef]
3. Chen, C.; Li, X.R. Fault diagnosis method of circuitbreaker operating mechanism based on coil current analysis. *J. Zhejiang Univ.* **2017**, *50*, 527–535.
4. Wang, Y.; Chang, G.; Han, K.; Qian, X.; Bao, Z.; Sheng, D. Parameter Design of a Self-Generated Power Current Transformer of an Intelligent Miniature Circuit Breaker Based on COMSOL. *Electronics* **2024**, *13*, 2285. [CrossRef]
5. Sun, S.G.; Zhang, Q.; Du, T.H.; Wang, J.Q.; Wang, Y. Study of evaluation method for low voltage conventional circuit breaker switching fault degree based on vibration signal. *Proc. CSEE* **2017**, *37*, 5473–5482.
6. Liu, H.L.; Xu, W.J.; Zhao, S.T. State Identification Method of Circuit Breaker Operating Mechanism Based on Time Course Waveform and Adaptive Spectrum Fusion Analysis of Vibration Signal. *High Volt. Eng.* **2022**, *40*, 3321–3329.
7. Yang, Y.W.; Guan, Y.G.; Chen, S.G.; Wang, J.; Zhao, K. Mechanical fault diagnosis method of high voltage circuit breaker based on sound signal. *Proc. CSEE* **2018**, *38*, 6730–6737.
8. Ledesma, P.; Jafary, P.; Repo, S.; Álvarez, A.; Ramos, F.; Della Giustina, D.; Dedè, A. Event-Based Simulation of a Decentralized Protection System Based on Secured GOOSE Messages. *Energies* **2020**, *13*, 3250. [CrossRef]
9. Yang, Q.; Ruan, J.; Zhuang, Z.; Huang, D. Chaotic Analysis and Feature Extraction of Vibration Signals From Power Circuit Breakers. *IEEE Trans. Power Deliv.* **2019**, *35*, 1124–1135. [CrossRef]

10. Li, X.; Zhu, K.; Guo, Z.; Zhang, Y.; Jia, S.; Jiang, X. Experimental study on arc interruption characteristics of SF6 and its mixtures with CF4. *Proc. CSEE* **2017**, *37*, 3315–3322.
11. Razi-Kazemi, A.A.; Niayesh, K.; Nilchi, R. A Probabilistic Model-Aided Failure Prediction Approach for Spring-Type Operating Mechanism of High-Voltage Circuit Breakers. *IEEE Trans. Power Deliv.* **2019**, *34*, 1280–1290. [CrossRef]
12. Zhao, S.; Zhu, J.; Wu, C.; Wang, B.; He, M. A correctiontesting method for mechanical characteristic parameter of circuit breaker basedon visiontechnology. *IEEE Trans. Electr. Electron.* **2018**, *13*, 1087–1091. [CrossRef]
13. Wang, L.; Tuo, Z.; Liu, G.; Zhang, J. Diagnostics on mechanical characteristics of high voltage circuit breaker based on machine vision technology. *High Volt. Eng.* **2020**, *46*, 2148–2154.
14. Bian, C.; Liu, S.; Xing, H.; Jia, Y. Research on fault-tolerant operation strategy of rectifier of square wave motor in wind power system. *China Electrotech. Soc. Trans. Electr. Mach. Syst.* **2021**, *5*, 62–69. [CrossRef]
15. Zhao, S.; Xu, W.; Li, Y. Identification methodfor energy storage state of spring mechanism of vacuum circuit breaker based on optimal multi-characteristics. *High Volt. Eng.* **2021**, *47*, 3777–3784.
16. Chen, X.C.; Feng, D.; Lin, S. Mechanical fault diagnosismethod of high voltage circuit breaker operating mechanism based ondeep auto-encoder network. *High Volt. Eng.* **2020**, *46*, 3080–3088.
17. van der Blij, N.H.; Purgat, P.; Soeiro, T.B.; Ramirez-Elizondo, L.M.; Spaan, M.T.J.; Bauer, P. Decentralized Plug-and-Play Protection Scheme for Low Voltage DC Grids. *Energies* **2020**, *13*, 3167. [CrossRef]
18. Zhao, S.; Wu, C.; Li, M. Research of circuit breaker mechanical characteristic test method based on normalized cross correlationpyramid-sector optimization algorithm. *Proc. CSEE* **2017**, *37*, 4265–4271.
19. Zhao, P.; Wang, J.; Xia, H.; He, W. A novel industrial magnetically enhanced hydrogen production electrolyzer and effect of magnetic field configuration. *Appl. Energy* **2024**, *367*, 123402. [CrossRef]
20. Wan, S.; Ma, X.; Chen, L.; Dou, L.; Lv, P.; Zhou, G. State evaluation and faultdiagnosis of high-voltage circuit breaker based on short-time energyentropy ratio of vibration signal and DTW. *High Volt. Eng.* **2020**, *46*, 4249–4257.
21. Cui, H.; Xia, S.; Zhou, K.; Zhang, X.; Zhang, M.; Sun, Y. Reliability prediction of 220 kV circuit breakers based on moffat rest hidden Markov degrada-tion process. *High Volt. Eng.* **2021**, *47*, 2108–2116.
22. Liu, H.L.; Xu, W.J.; Zhao, S.T.; Jiang, C.Y.; Liu, J.M. Feature extractionmethod of operating mechanism state of high voltage circuit breakerbased on energy trajectory entropy. *High Volt. Eng.* **2022**, *48*, 2945–2951.

energies

MDPI

Article

A Novel Concept of High-Voltage Balancing on Series-Connected Transistors for Use in High-Speed Instrumentation

Alexandr Despotuli *, Viacheslav Kazmiruk, Anastasia Despotuli and Alexandra Andreeva

Institute of Microelectronics Technology and High Purity Materials, Russian Academy of Sciences (IMT RAS), 142432 Chernogolovka, Russia; kazmiruk@iptm.ru (V.K.); despotuli.aa@iptm.ru (A.D.); andreeva@iptm.ru (A.A.)
* Correspondence: despot@iptm.ru

Abstract: The novel concept of reliable voltage balancing on N fast high-voltage (HV) transistors, connected in series, is verified by computer modeling/experimental testing. The essence of the concept is to transfer the balancing function from conventional snubbers, resistive dividers, varistors, etc., or sophisticated gate-side control techniques, to "individual" resistive loads (of transistors) connected to "individual" HV sources of power. The concept has been implemented in the recently patented architecture of HV rectangular pulse generators. The operation of any series-connected stack requires (1) synchronization of control actions on gates of all N transistors; (2) static HV balancing on all transistors in OFF states; and (3) dynamic HV balancing during ON↔OFF transients. The goals of the new design are to achieve an exceptionally high level of HV balancing in modes (2) and (3), as well as to simplify the process of configuring/customizing the circuit. Testing confirms that new generators exhibit minimal ripple during ON→OFF transients. Reliable operation with high-quality rectangular pulses is ensured even at a voltage slew rate of more than 100 kV/μs, while each transistor blocks a voltage close to the maximum value specified in its datasheet. The presented novelties are likely suitable for high-speed instrumentation.

Keywords: high-voltage; rectangular pulses; series-connected transistors; voltage balancing; high-voltage generators; high-speed instrumentation

1. Introduction

High-voltage (HV) rectangular pulse generators with different architectures, designs, circuit topologies, and switch types [1–5] find applications in a wide variety of electronic systems used in scientific experiments, national defense, biomedical treatments, tests, and measurements. Wide-gap semiconductor switches have excellent performance parameters. Silicon carbide (SiC) transistors are superior to their silicon counterparts in a number of aspects [6,7]: electrical breakdown field, thermal conductivity, saturated electron drift velocity, on-resistance, and some others.

For a field-effect transistor (FET), the gate length determines the speed characteristics and the time required for electrons to travel through the gate-driven channel. Therefore, in a comparison of different FETs, it can be expected that the switching speed of HV transistors (different manufacturers [8–14]) decreases as the maximum blocking voltage (U_{max}) and on-state current rating increase. Although these trends are possible to visualize through SPICE modeling, it is not always possible to directly compare the switching characteristics using the same computer software. There are many specialized software programs, but some of them model transistors from specific manufacturers, while using SPICE models from external sources can lead to problems and errors.

The highest U_{max} of commercial SiC transistors is still below 5 kV. Examples include 3.3 kV [12] and 4.7 kV [13]. Identical HV transistors connected in series can provide a combination of high switching speed between OFF and ON states and high blocking voltage ($U \sim N^{*}U_{max}$). In stacks, series-connected transistors also demonstrate (1) a lower drain-to-source resistance in ON state, (2) a higher current density, and (3) are less susceptible to the effects of cosmic radiation compared to a single HV transistor [15]. Stacks have a wide range of applications in the area of power electronics and converter systems. HV rectangular pulse generators belong to the mentioned areas and, in addition, fall into the subclass of high-speed instrumentation [16,17].

For an ideally designed stack, the slew rate of a rectangular HV pulse depends on the speed characteristics of wide-gap HV transistors that become faster and faster over time. **Therefore, the design of stacks, reliable, easily, and flexibly adapted to any type of high-speed transistors, is an urgent task.**

Real transistors exhibit variations in parameters; this involves well-known technical problems related to voltage imbalance on transistors [3–5]. The perfect balance is particularly difficult to achieve in cases of high-quality HV rectangular pulses, as quasi-rectangular waveforms have a large voltage slew rate, dU/dt. When operating in such challenging conditions, the first transistor performing the ON→OFF transition (and the last transistor performing the OFF→ON transition) receives most of the stack voltage; this can cause the transistor to fail.

HV pulse generators exist in two basic configurations: (1) S-L-St-G and (2) S-St-L-G, where St is the stack of N transistors, L is the resistive load, S is the HV source of power, and G is the ground node. Here, we deal with the S-L-St-G configuration, in which the decoupling capacitor eliminates the constant HV on the object of influence [18].

There are many ways to improve the quality of HV equalization (balancing) on series-connected transistors of the same type. The combination of advantages and limitations of particular techniques is analyzed in a number of reviews [1–5,15]. For instance, in the large-scale production of HV pulse generators [18,19], specialized companies may employ pre-selected transistors that are not exactly identical but vary only slightly in terms of parameters. However, this approach to voltage equalization requires a large number of transistors for per-selection. This is one of the current methods for solving the voltage equalization problem.

In sophisticated gate-side methods of voltage equalization [20], high-speed control circuits respond to HV in N×MOSFET and regulate the gates of all transistors. The complexity of setting up the entire system where the stack is in the states with large voltage slew rates on the order of 100 kV/µs limits the use of such feedback.

In practice, common approaches to voltage equalization across transistors of the same type include the use of resistive dividers, varistors, and snubbers to prevent voltage imbalance [3,21,22]. These approaches also have well-known limitations [3,22]. In particular, there is a lack of universality in using specific chains of electronic components that respond to a certain frequency range (for example, the snubber time constant [22]). The efficiency of known dynamic equalization methods decreases with increasing switching speed. The complexity of the problem increases significantly if the stack contains fast transistors of different types.

Some reviews [23,24] describe advanced methods for voltage equalization. We developed a new concept for voltage balancing for stacks with all transistors of different types. Our concept leads to the creation of innovative circuitry for HV pulse generators suited for high-speed instrumentation [25,26]. New generators enabled the pulsed vacuum emission of ions (electrons) from liquid metals [25]. Such generators will find applications in our R&D for advanced superionic conductors and emission nanoionics.

A key feature of the new HV generators is the use of several (N) independent HV sources of power and, correspondingly, N resistive transistor loads (as opposed to one source and one load in traditional designs [3], see Fig. 9 in [17]). These innovations provide reliable, fast, and high-quality voltage balancing in serial-connected transistors.

It is important to differentiate between the term "voltage equalization", which is appropriate for stacks containing transistors of the same type, and "voltage balancing", which more accurately describes the technical effect related to the proposed concept. Thanks to the proposed new architecture of HV pulse generators, transistors with similar dynamic characteristics but significantly different U_{max} can function consistently in a stack. In this context, the term "voltage equalization" refers to one particular case of voltage balancing.

The modern various SPICE software allow one to clearly prove the correctness, functionality, and usefulness of a particular circuit architecture, e.g., HV pulse generators. However, when it comes to implementing the HV circuit design in a printed circuit board (PCB) layout, some important factors for concrete customers, such as electromagnetic compatibility or the interfering influence of parasitic components, still need to be minimized.

In this study, we investigate the applicability limits of the proposed voltage balancing concept [25,26] by means of circuitry modeling and experimental testing. We use an original methodology of combining transistors of different types in a stack. The content of the concept becomes clear with examples of the operation of new HV generators. We identify inherent trade-offs and disadvantages of proposed innovation. Due to its simplicity, adaptability, and operational efficiency, the new technical solution has the features of basic circuits of power and HV electronics. The innovation has potential applications in the area of high-speed instrumentation.

2. Novel Architecture of HV Pulse Generators

The flowchart of HV pulse generators includes three main units depicted as I, II, and III in Figure 1. Unit I consists of N transistors (1.1–1.N) connected by N-1 electrical contact (2.1–2.N-1) sequences. Unit II consists of N-regulated HV sources of power (4.1–4.N) shunted by N capacitors (5.1–5.N). Unit III controls the gates of transistors in N × MOSFET (N × IGBT or devices with cascode configuration [14]). N resistive loads (i.e., 3.1–3.N) of the N transistors (1.1–1.N) connect Units I and II.

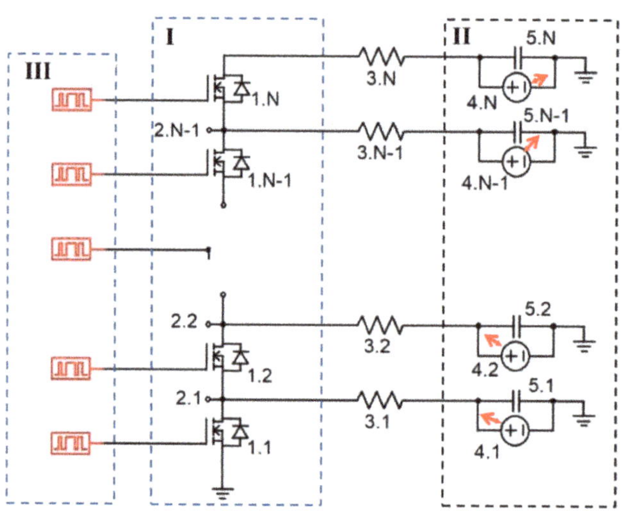

Figure 1. The patented design of HV pulse generators [25].

A thorough analysis of the current state-of-the-art basic circuits and their variations (for example, [1–5,20–24]) reveals the innovative and surprisingly simple design of HV generators proposed in [25,26]. The new architecture does not solve the issue of synchronous operation on transistor gates in Unit I, nor does it resolve the challenges of a specific implementation of Unit II.

Looking ahead, we note that experimental new generators with low-cost HV SiC MOSFETs provide switching with voltage slew rates of more than 100 kV/μs for quasi-rectangular pulses at an effective repetition rate of about 7 MHz. SPICE modeling shows that it is possible to achieve voltage slew rates of about 500 kV/μs for modern HV transistors.

3. Voltage/Current Distributions in Circuits of New Generators

In new generators, the mechanisms that ensure normal operation of HV transistors in the serial stack manifest themselves differently in steady-state and transient modes. Modeling of the circuit behavior allows one to find out the characteristics of switching in specific circuit architectures. The solutions of Kirchhoff equations for the electrical equivalent circuits (Figures 2 and 3) indicate the process parameters for OFF and ON steady-states.

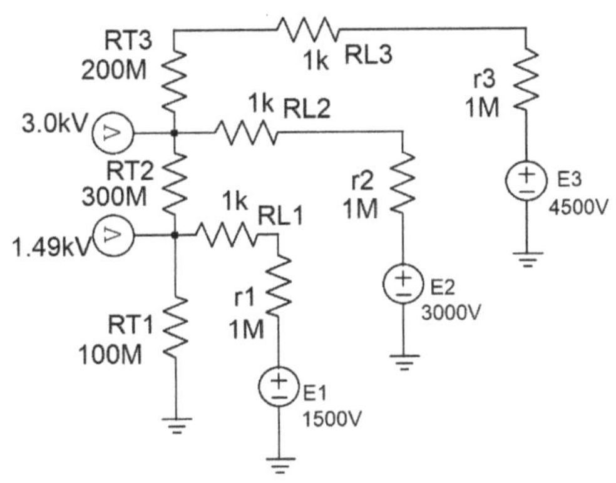

Figure 2. Voltages in the equivalent circuit of new generators with 3×MOSFET stack in OFF state.

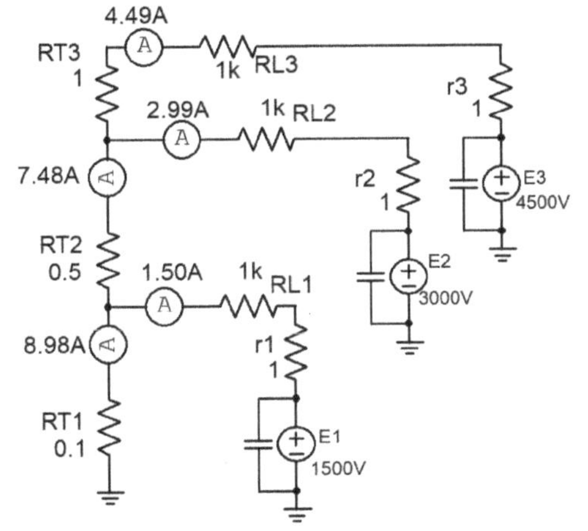

Figure 3. The distribution of currents in the equivalent circuit of new generators with 3×MOSFET stack in ON steady-state.

3.1. Equivalent Circuit with 3×MOSFET in OFF Steady-State

For the OFF state of new generators, the main interest is the voltage distribution on transistors. Figure 2 shows the equivalent circuit of a generator in OFF state. The sources of regulated HV power (E1, E2, and E3) have effective internal resistances, depicted as r1, r2, and r3. These resistors may be about 1 MΩ. In these conditions, the E sources with voltages of about 5 kV can produce a constant current of ~5 mA. Datasheets typically provide zero gate voltage drain currents (I_{DSS}) for U_{max}, allowing for the estimation of transistor resistances (RT1, RT2, and RT3) in OFF state. For example, STP4N150 transistors (U_{max} = 1.5 kV) have $I_{DSS} = 10^{-5}$ A [9], and OFF-resistors are ~10^8 Ω.

Multisim software (of 14.3 version) simulations show a high quality of voltage equalization/balancing when the stack is in OFF state: all transistors (T1, T2, and T3) with a large variation of resistance values (RT) appear to be under the same voltage.

3.2. Equivalent Circuit for New Generators in ON Steady-State

In ON state, it is necessary to maintain appropriate currents on transistors in order not to overload them. The position of Ti transistor in N×MOSFET relative to the ground node (G) determines the current (ITi) of Ti transistor. The closer the Ti transistor is to G-node, the more that current flows through it.

In HV SiC transistors, typical ON state resistances range from 0.1 to 10 Ohm. This is much lower than the transistor loads. HV sources have low effective impedances because large capacitors are connected to them in parallel. The latter provide significant pulse currents through the open channel of transistors. Figure 3 shows the results of the electrical circuit simulation for ON steady-state (dc current) of 3×MOSFET.

The simulation results show that the currents (ITi) flowing through the transistors in 3×MOSFETs are:

$$\mathrm{IT3} = \mathrm{E3/RL3},\ \mathrm{IT2} = \mathrm{IT3} + \mathrm{E2/RL2},\ \mathrm{IT1} = \mathrm{E3/RL3} + \mathrm{E2/RL2} + \mathrm{E1/RL1} \quad (1)$$

i.e., IT1 current is determined by summation, even if all transistors of the stack have different resistances in ON state.

3.3. Disadvantage of Proposed Concept

Let us assume that for all HV sources (Figure 3), the value of Ei is an integer multiple of E1 = E, where E represents the voltage difference on all transistors in a stack. If all load resistances, RL1, RL2, RL3, etc., satisfy the condition that they are equal RL, then the current IT1 through the T1 node (1) will be:

$$\mathrm{IT1} = (3\mathrm{E} + 2\mathrm{E} + \mathrm{E})/\mathrm{RL} \quad (2)$$

For a stack with identical N transistors and identical loads, Equation (2) takes the form of

$$\mathrm{IT1} = (\mathrm{E/RL})\,(N + \ldots + 2 + 1) = (\mathrm{E/RL})\,(N + 1)*(N/2) \quad (3)$$

On the contrary, for a conventional stack (with a single load RL) under the corresponding voltage $N*$E, the current IT1 is $N*$E/RL, i.e., it is $(N + 1)/2$ times smaller than in Equation (3). Multipliers of order $(N + 1)/2$ are a disadvantage of the new technical solution. Therefore, the application area of the proposed concept is most likely not industrial electronics. The field of high-speed instrumental applications with their relatively low operational currents and significant demands on the quality of functional performance is promising.

3.4. Equivalent Circuit in Steady-State at Desynchronization of Transistor Switching

One of the beneficial aspects of the concept is the decreased voltage overload of the T1 transistor, which delays ON→OFF transition compared to other transistors in the stack. This is because each transistor has its own HV power source and load resistance.

Let us consider the case of 2×MOSFET stack in which the T2 transistor is in ON stationary state while T1 with U_{max} = 1.5 kV is still in OFF state. The voltage overload on transistor T1 is shown in Figure 4. This overload is 2.25 kV − 1.5 kV = 0.75 kV but not 3 kV − 1.5 kV = 1.5 kV, as it would be in a conventional generator.

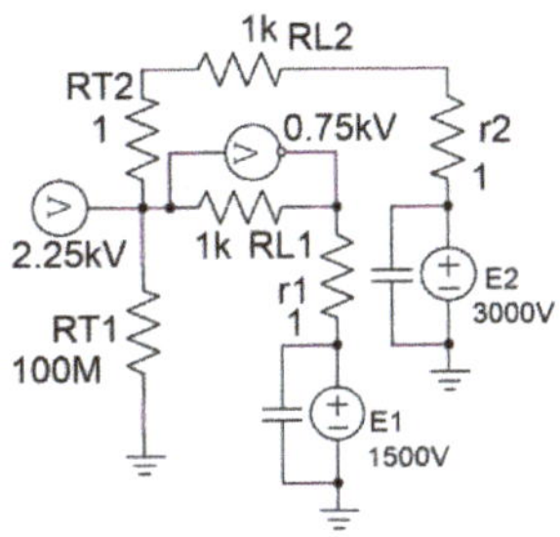

Figure 4. The reduction of overvoltage on the T1 transistor (from 1500 V to 748 V) is due to the corresponding voltage drop on RL1 load.

The reduction in the overload results in a voltage drop across RL1 load of the T1 transistor. Therefore, in this specific case, the overload for the T1 transistor, which lags the transition from OFF to ON, is half the voltage it would be in a standard circuit with a single total load.

3.5. Equivalent Circuit at Desynchronization in Transient Switching Mode

Below, we demonstrate the high quality of voltage balancing across stack transistors during transient switching modes. The presented data relate to test experiments and circuit modeling using SPICE models of modern HV transistors. As the load resistance increases, ON→OFF transition time also increases. To verify the proposed concept in dynamic modes, we need transistors capable of fast switching between states. Information on the required transistors can be found in datasheets and by using computer SPICE models at transient circuit simulation.

The time delay for the t transistor relative to the T2 in OFF→ON transient mode (circuit as in Figure 4) is shown in Figure 5 for E1 = 1000 V and E2 = 2000 V. The Micro-Cap program simulates the operation of 2×STP4N150 circuit with RL1 and RL2 being 400 Ω loads. In new generators, transistor T1 is under less voltage overload than what can occur in conventional generators.

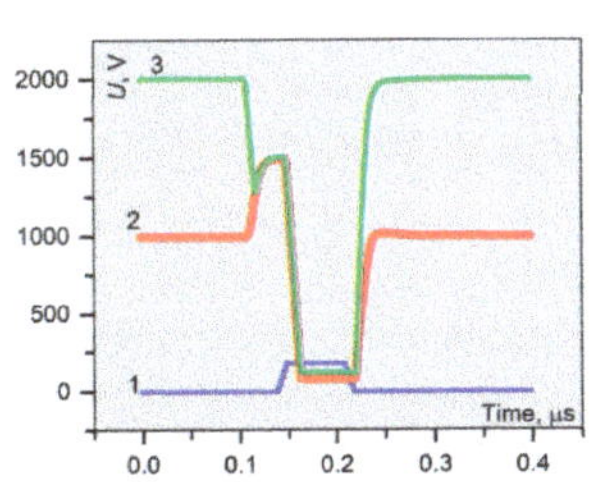

Figure 5. Overload reduction from 1000 V to 500 V for the T1 transistor in 2×STP4N150 stack, which operates in dynamic mode (the scheme for modeling is as in Figure 4). 1 (blue): gate voltage waveform T1 (10× magnification); 2 (red): drain voltage waveform T1; 3 (green): drain voltage waveform T2.

4. Modeling/Experimental Testing for Verification of Proposed Concept

To convincingly validate the proposed concept of voltage balancing, HV transistors with fast switching must be involved in modeling/experimental testing. This implies a pre-selection among different types of MOSFETs that are faster than IGBT devices.

For HV rectangular pulse generators, the design of PCB is such a complex task that it is necessary to take into account small stray capacitances and inductances to a greater extent than in the case of low-voltage (LV) electronics. If PCB is not properly designed, undesirable effects can occur in the circuit [27], which will distort HV pulses. For example, if the pulse amplitude is 10 kV and the pulse rise time is about 10 ns, then the parasitic capacitance of 10 pF on TN transistor drain creates a current of about 5 A through the stack. Such a value is comparable to the working pulse current. In these conditions, parasitic capacitance can distort the shape of a quasi-rectangular pulse due to the overcurrent of the T1 transistor. To the best of the authors' knowledge, among the many PCB design software, only Altium [28] (version 25) checks PCB layouts against HV design rules.

Under these conditions, it is difficult to properly and comprehensively verify the new circuit concept in experimental tests for different types of fast transistors. Therefore, this study focuses on modeling different variants of schemes. The circuit simulation programs (such as Multisim: v. 14.3, Micro Cap: v. 12.2, LTspice: v. 24, QSpice: v. 08/16/2024, and PSIM: v. 2024.0.0247) provided tests of switching for HV SiC transistors. We also present experimental test data. For instance, experimental generators with wire connections between circuit elements in their design achieved high switching speeds with a high repetition rate.

4.1. Pre-Selection of Fast Transistors

Figures 6 and 7 show the results of the comparative modeling (LTspice). Only waveforms of SiC transistors STP4N150 (STMicroelectronics, [9]), SCT1000N170 (STMicroelectronics) and G2R1000MT33J (GeneSiC Semiconductor [12]) are indicated, as they were selected for experimental verification of the proposed voltage-balancing concept.

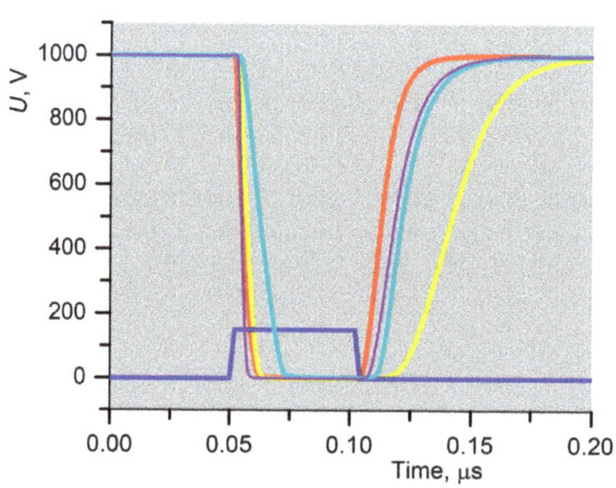

Figure 6. The simulated switching waveforms of some modern SiC transistors (different manufacturers). The voltage waveforms for gates (blue, 10× magnification) as in Figure 5. Red: the voltage on the drain of SCT1000N170 transistor, selected for further tastings. Other transistors are not mentioned for ethical reasons (they were not selected for further comparative testing).

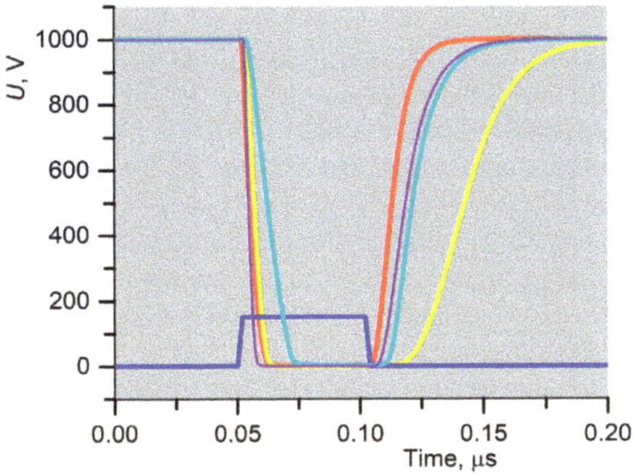

Figure 7. The comparison of simulated switching waveforms of some modern SiC transistors (different manufacturers). Pulses on the gates of transistors (blue, 10×), as in Figures 5 and 6. Red: the voltage on SCT1000N170 drain; yellow: the voltage on STP4N150 drain. Other transistors are not mentioned for ethical reasons.

The GeneSiC [12] offers encrypted models of transistors that are only suitable for LTspice/PLECS programs, while the new and growing Qspice program [14] currently contains cascode models and only a few other HV transistors in its library. The selected transistors have the following characteristics: STP4N150, which is the low-cost and high-switching-speed transistor: U_{max} = 1.5 kV, pulsed drain current 12 A; SCT1000N170: U_{max} = 1.7 kV, pulsed drain current 20 A; and G2R1000MT33J: U_{max} = 3.3 kV, pulsed drain current 10 A.

Figures 6 and 7 demonstrate that the switching of the selected HV transistors do not differ significantly in terms of rise and fall times, as well as turn-on (and turn-off) delay times. According to the simulation (Figure 8), the slew rate of STP4N150 at turn-on is approximately equal to the same value for the SCT1000N170 transistor; the G2R1000MT33J is a better choice, while the HV IXTT1N450HV ([13]) transistor (U_{max} = 4.5 kV) shows a noticeably lower switching speed (it can be suitable for an operation at the microsecond scale).

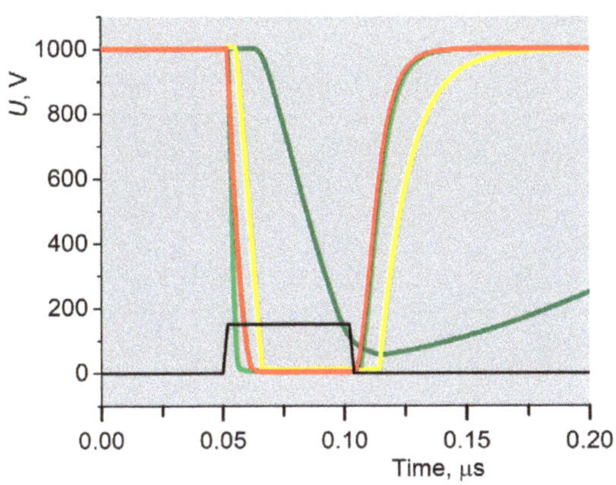

Figure 8. The simulated switching voltage waveforms for transistors: SCT1000N170 (red); STP4N150 (yellow); G2R1000MT33J (green); and IXTT1N450HV (dark green). Pulses on the gates of transistors (black, 10×), as in Figures 5–7.

4.2. Pre-Selection of Transistors for Experimental Stacks: New Methodology

If the proposed voltage balancing concept was tested by simulating the operation of a stack with identical transistors, the verification methodology would be considered questionable. Indeed, for identical transistors, all SPICE models are identical, and if real transistors had identical parameters, voltage balancing would not make sense at all.

Our original methodology for verifying the new concept of HV balancing in series-connected fast transistors involves modeling the behavior of stacks with transistors that are not of the same but different types. Balancing the stacks during the transitions between OFF and ON is difficult using conventional techniques based on resistors, snubber circuits, varistors, and other similar components. This is true for techniques that use gate-side controls, which need to take into account different transistor types and operating modes. In this context, the new concept has no limitations; it seems quite simple, universal, and reliable.

Figures 9 and 10 show the simulation results in LTspice (v. 24) for two new generators with different stacks. Figure 9 corresponds to the case when one SCT10000N170 transistor is on the T1 site and G2R1000MT33J is on the T2 site. In Figure 10, two SCT10000N170 transistors (connected in parallel) are on the T1 site, while G2R1000MT33J is on the T2 site (the circuit is shown in Figure 11).

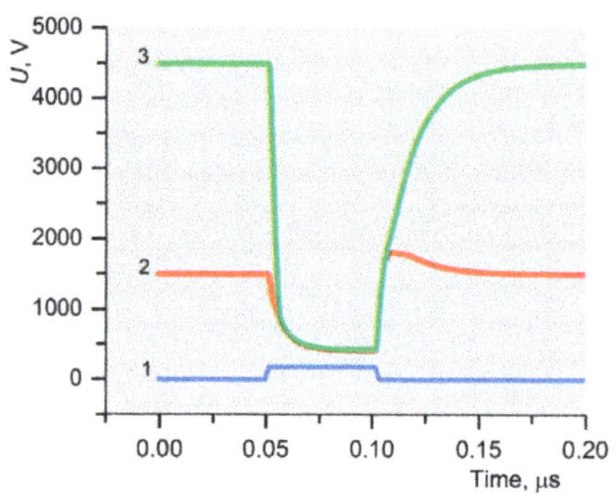

Figure 9. The simulation-switching voltage waveforms for the new generator with 2×MOSFET stack. 1: pulses on the gates of transistors (blue, 10×); 2: SCT1000N170 transistor at the T1 site (red); 3: G2R1000MT33J transistor at the T2 site (green).

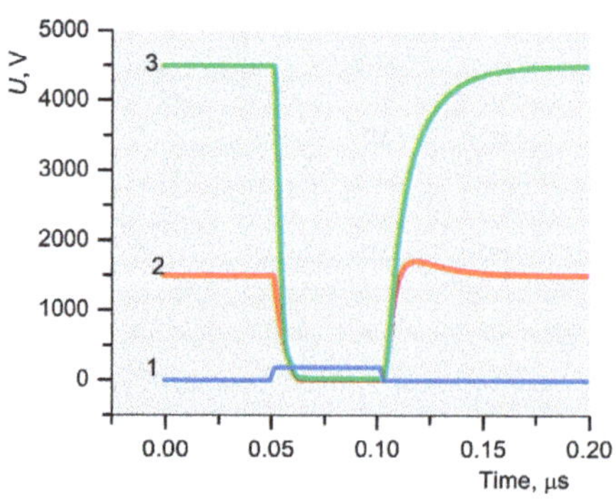

Figure 10. The simulation stitching for the new generator with a hybrid stack. 1 (blue): voltage pulses on gates of transistors (10×); 2 (red): the voltage waveform on drains of two parallel-connected SCT1000N170 transistors at the T1 site; 3 (green): the voltage waveform on the drain of the G2R1000MT33J transistor at the T2 site.

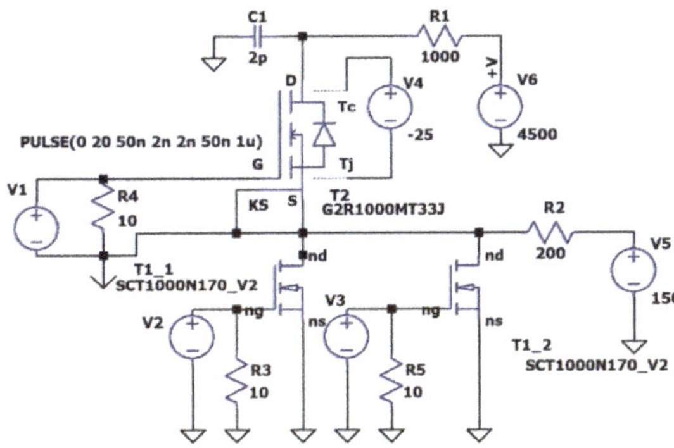

Figure 11. The circuit of the new generator with the hybrid dual-cascade stack (two SCT10000N170 transistors at the T1 site and one G2R1000MT33J transistor at the T2 site).

A comparison of Figures 9 and 10 captures a disadvantage of new generators: the closer the transistor is to the ground node (G), the more current flows through it. To overcome this feature, we need, for example, to put two transistors at the T1 site. This will reduce the average current amplitude through each transistor, increase the pulse amplitude, and improve the shape of quasi-rectangular HV pulses. Note that if the number or type of transistors changes at the T site, new schematics enable flexible customization of the stack by adjusting resistive loads.

Component C1 (Figure 11) symbolizes the unavoidable parasitic capacitance; it strongly affects the current flowing in high-speed switching stacks. Hence, the issue of PCB design for rectangular HV pulse generators is topical.

To show how effective the proposed concept is, we have simulated the drain voltages for a stack consisting of three very different series-connected HV transistors. The results are shown in Figure 12.

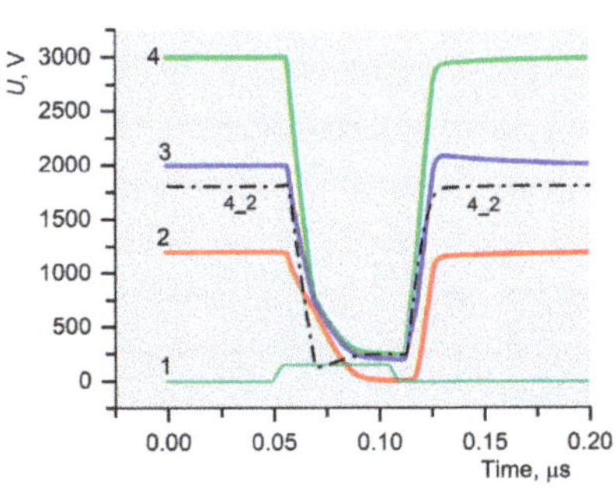

Figure 12. Voltage balancing on series-connected transistors of different types in the new generator. 1 (thin dark green): Pulses on gates of all transistors (10×); 2 (red): the voltage waveform for the drain of UF3C170400K3S cascode transistor at the T1 site; 3 (blue): the voltage waveform for the drain of STP4N150 at the T2 site; 4 (bold light green): the voltage waveform for the drain of STP4N150 at the T3 site (green). The 4_2 sign (black, dash-dotted) denotes the difference between T3 and T1 voltage waveforms (Qspice simulations).

The UF3C170400K3S transistor is a composite device in which a normally-on SiC JFET is combined with Si-MOSFET to produce a normally-off SiC-FET hybrid. In the modeled stack operation, STP4N150 transistors are on T2 and T3 sites. Note: due to the different currents on the T2 and T3 sites, the corresponding same-type transistors have different resistance at the identical voltage difference, i.e., they are in different states.

4.3. Balance Processes in Conventional Generators vs. in Proposed Concept

Let us examine in more detail the scenarios of processes involved in voltage equalization/balancing within a stack, in which transistors (T1, T2, . . ., TN) may exhibit differences in characteristics. Two possible cases are considered: (•) a single common load RL serves the entire stack; (••) "individual" loads RLi serve each transistor Ti of the stack (according to the proposed concept).

The case of OFF→ON transitions.

(•) During the transition (most of the time t), any Ti transistor in the stack has a resistance RTi(t) significantly larger than RL. Therefore, in the specified time interval, the voltage from Ti is redistributed not to RL load but to transistors in the stack that are at earlier stages of OFF→ON transition. This process can cause the stack to break down.

(••) If each stack transistor has its own individual RLi load, then for a long period of time during the transient process, each resistance RTi(t) satisfies the ratio: RTi(t)~RTk(t) >> RLi. In this case, transient currents flow through all loads, and the voltages are redistributed between the corresponding pairs (Ti and RLi). Such localized processes do not significantly affect each other, even if transistors block different voltages in the OFF state.

The case of ON→OFF transitions.

(•) In a stack with a single RL load, for any Ti and Tk transistors, RTi(t)~RTk(t) >> RL during a long part of the transition time. Therefore, the voltage redistribution occurs during the process of freeing the devices from excess charge carriers, when charges pass through the entire sequence of transistors.

(••) If each Ti transistor in the stack has an individual load RLi << RTk(t), then the Ti transistor will be released from excess charge via RLi. This process is faster than transferring charge through a sequence of megaohm transistors, as in the case of a single common load (RL).

Thus, if all stack transistors have their own "individual" loads, then the processes of voltage redistribution in dynamic modes follow less resistive paths. The disadvantage of the proposed new solution is increased currents on transistors with small "i" indices, e.g., T1. Transistor parallelism partially solves the problem. Another disadvantage is an enhanced power consumption of multi load stacks.

5. Experimental Generators: Developing and Testing

5.1. Generator Based on 2×STP4N150 with U_{max} = 1500 V

According to preliminary computer simulation results, the STP4N150 and G2R1000MT33J transistors took part in test experiments to obtain data confirming the operational advantages of the proposed voltage-balancing concept. The quality of voltage balancing across stack transistors decreases significantly with increasing voltage slew rate. Therefore, it is important that new generators can operate reliably under transistor-switching challenging conditions.

Figure 13 shows the generator with 2×STP4N150 and two HV drives of ADUM4223 type (Unit III). HV isolation separates the input low voltage (LV) and output HV sections of the drivers. The monolithic transformers transmit signals between the LV and HV sections in ADUM4223 chips. Autonomous miniature 16 V sources power the HV section of drivers. We pre-selected drivers based on the results of signal propagation delay measurements. This allowed the drivers to synchronously drive the gates of STP4N150 SiC MOSFETs.

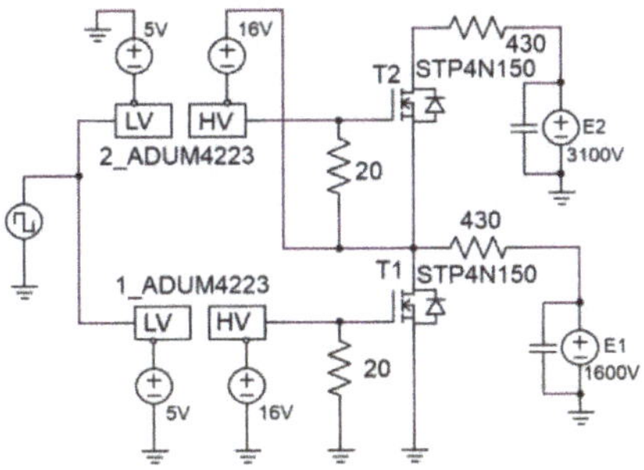

Figure 13. The new generator with two STP4N150 transistors (rated at 12A drain pulsed current and U_{max} = 1500 V).

With two selected ADUM4223 drivers or two selected HV UCC21520 drivers, new generators produce single, high-quality rectangular HV pulses. Moreover, owing to high-quality HV equalization, ADUM4223 chips have demonstrated the ability to drive the gates of STP4N150 transistors at an effective frequency of more than 7 MHz. This is seven times the maximum data rate with a minimum pulse width of 50 ns specified in the datasheet for ADUM4223 chip.

Figure 14 shows voltage waveforms for the generator with 2×STP4N150, where pulses are initiated by two ADUM4223 drivers. In the OFF state of both STP4N150 transistors, it was possible to set supply voltages that are slightly (50–100 V) higher than their U_{max}. In other words, during the switching processes, the voltage on the 2×STP4N150 was slightly higher than the maximum allowable voltage for the stack (extremely challenging transient test). Modeling in Micro-Cap 12 confirmed the experimental data obtained. The model of the new generator in the OFF state used E1 = 1600 V and E2 = 3100 V on the T1 and T2 transistors, respectively.

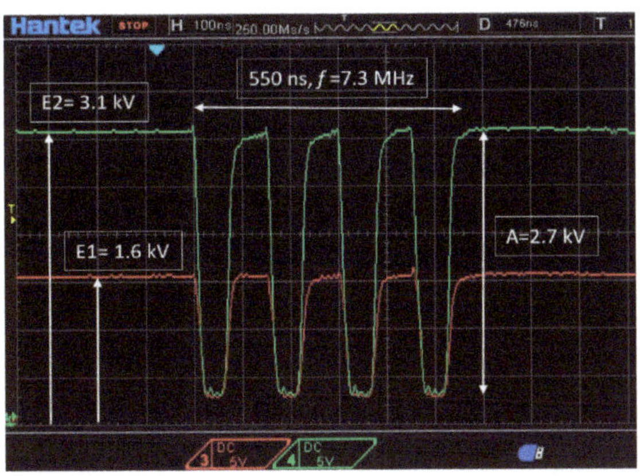

Figure 14. Voltage waveforms on drains of two STP4N150 transistors in the new generator (experiment). Scales are 500 V/div and 100 ns/div. HV probes are 1:100. Red: the waveform for the drain of STP4N150 at the T1 site. Green: the waveform for the drain of STP4N150 at the T2 site.

We also confirmed the switching stability of the circuit (Figure 13) under the following conditions: two STP4N150 transistors (connected in parallel, rated at 24A total pulsed drain current) or one STW9N150 transistor (rated at 32A pulsed drain current) operated at the T1 site. It is important to note that after replacing the MOSFETs, no circuit reconfiguration was required. The observed situation is very different from that expected for other voltage equalization techniques using resistive dividers, snubbers, varistors, etc.

5.2. Two Generators Based on G2R1000MT33J Transistors with Umax = 3300 V

Figure 15a,b illustrates the operation (experimental test and corresponding LTspice simulations) for two generators, in which stacks contain the G2R1000MT33J transistors (rated for a pulsed drain current of 10 A) at the T1 and T2 sites. In the case of Figure 15a, the circuit was as it is depicted in Figure 13, with two G2R1000MT33J transistors. The resistive loads were R1 = 860 Ω and R2 = 860 Ω, while E1 = 2.6 kV and E2 = 5.2 kV. The quality of the rectangular pulses depends on the current at the T1 site. The simulation shows, in Figure 15b, that if two transistors are connected in parallel at the T1 site, the amplitude and quality of the rectangular pulses significantly increase.

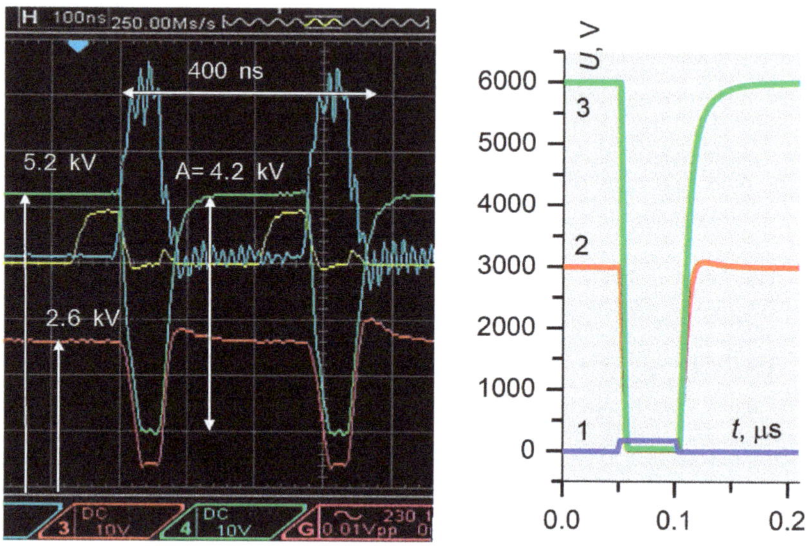

(**a**) (**b**)

Figure 15. Verification of operational efficiency of new generators. (**a**) Experimental voltage waveforms on the drains of G2R1000MT33J transistors in two cascade stacks of new generators (the circuit as in Figure 13 with R1 and R2 loads of 860 Ω). Scales are 1 kV/div and 100 ns/div. HV probes are 1:100. Red waveform: the voltage on the transistor drain at the T1 site. Green waveform: the voltage on the transistor drain at the T2 site. Yellow waveform: the input signal for both ADUM4223 drivers. Cyan waveform: the voltage on the output of 1_ADUM4223 driver. (**b**) Simulated voltage waveforms on the drains of G2R1000MT33J transistors in two cascade stacks with two paralleled transistors at the T1 site (the circuit as in Figure 11 with loads R1 = 200 Ω and R2 = 1 k Ω). Scales are 1000 V/div and 100 ns/div. 1: pulses on gates of two paralleled transistors at the T1 site (blue, 10× magnification); 2: the waveform for the combined drain of two G2R1000MT33J transistors at the T1 site (red); 3: the waveform for the drain of G2R1000MT33J at the T2 site (green).

By using STP4N150 and G2R1000MT33J transistors, the series-connected stacks were able to generate quasi-rectangular pulses with peak voltages of 2.7 kV and 4.2 kV. In these cases, slew rates greater than 100 kV/μs were attained at turn-on and turn-off processes (for comparison, see [29]). For 2×STP4N150 stack, the effective frequency was higher than 7 MHz. The results of HV waveform modeling showed that the switching speed for hybrid series-parallel stacks with two cascades can reach a slew rate of 500 kV/μs. The proposed innovations demonstrate reliable performance in challenging conditions. An example is fast switching near the safe operating area (SOA) and $N \times U_{max}$ limit voltages applied to the stack.

Thus, as proven by modeling and experimental testing, the novel concept resulted in excellent HV balancing across the stacks in new HV rectangular pulse generators.

6. Conclusions

The design of stacks, easily and flexibly adapted to any type of high-speed transistors, is an urgent task. The technical result from the proposed concept of reliable voltage balancing on a stack of fast high-voltage (HV) transistors creates the following advantages:

1. The voltage on each transistor can be very close to the nominal maximum U_{max}. This ensures reliable operation, enhances the amplitude of HV pulses, minimizes the rise/fall time transients, and attains a high pulse frequency.
2. The new circuits are simple and can be easily customized/modified to meet different requirements. The circuits can contain transistors with different U_{max} ratings. Transistors in parallel increase the power of pulses without changing any other parts of the circuit.
3. The creation of new HV pulse generators does not demand knowledge of the subtleties of the functioning of snubbers, experience in the adjustment of series-connected stacks, or skills relating to sophisticated gate-side control techniques.

The main disadvantage of new HV generators is the summation of currents flowing through all transistor loads. The closer the T transistor of the stack is to a ground (G) node, the more current flows through it. Therefore, the application area of the proposed concept is most likely not industrial electronics but rather the field of high-speed instrumentation with its relatively low operating currents and significant quality requirements for functional performance.

In this work, we only briefly examined the consequences of current flowing through a grounded (G) transistor of a stack. This current is $(N + 1)/2$ times larger than in a traditional technical solution. Also, issues related to connecting additional HV sources, increasing electromagnetic interference, thermal management, and circuit design complexity in terms of controlling various parasitic components have not been investigated. The ability of the new generator to provide HV pulses coupled with high currents requires further research. This includes the important question of the frequency and power limit for the proposed new HV generator architecture.

GaN discrete FETs offer outstanding speed and power, and the advances will continue to expand the frequency range and increase power level. However, there are two points to consider. First, technological progress in extremely high frequency operation is achieved by reducing the gate length in GaN devices; this path is limited for HV FETs but not for power and HV electronics based on serial-connected transistors. Second, SiC still operates at higher voltages than GaN. Therefore, in this study, we have not systematically applied the concept of HV balancing to GaN FETs. Trial simulations (LTspice, v. 24) for hybrid series-parallel stacks using three GaN GPIHV30SB5L [10] transistors, as in Figure 11, demonstrate a high level of voltage balancing. Switching slew rates of this hybrid stack were higher compared to a similar stack with three SiC G2R1000MT33J transistors. A total blocking voltage of 2 kV was on both types of stacks.

Overcoming and eliminating the drawbacks associated with trade-offs is the inevitable fate of any technical solution, which is never universal but fulfills with guarantee its main target functions. The proposed innovations perform their core functions excellently and will facilitate new research on yet unexplored topics. The new conception of HV balancing and generator architecture are simple, versatile, and efficient in their main functions. For this reason, the innovation has the features of basic circuits of power and HV electronics, high-speed instrumentation (for examples, see [2,16,17]), and nanosecond biomedical treatments [30,31]. In fact, the proposed circuit can "extract" the ultimate speed characteristics from any transistor in a stack.

The significance of the proposed and verified HV balancing concept is useful to consider from different general points of view. Expert committees often have to address the relative scientific merit of findings. The article [32] proposes a quantitative framework for such deliberations, and it argues that the degree of surprise of an experimental result is an adequate metric for evaluating the scientific merit. This thesis is consistent with [33], that while the number of publications grows exponentially, the "cognitive extent of science" (the number of different unique concepts) increases only linearly.

Author Contributions: Conceptualization, A.D. (Alexandr Despotuli); methodology, A.D. (Alexandr Despotuli), V.K., A.D. (Anastasia Despotuli) and A.A.; experimental investigation, A.D. (Alexandr Despotuli) and A.D. (Anastasia Despotuli); software and computer modeling, A.D. (Alexandr Despotuli), A.D. (Anastasia Despotuli) and A.A.; resources, A.D. (Alexandr Despotuli), V.K. and A.A.; writing—original draft preparation, A.D. (Alexandr Despotuli) and A.D. (Anastasia Despotuli); writing—review and editing, A.D. (Alexandr Despotuli), V.K., A.D. (Anastasia Despotuli) and A.A. All authors have read and agreed to the published version of the manuscript.

Funding: This research was funded by STATE TASK No. 075-00295-25-00.

Data Availability Statement: The original contributions presented in the study are included in the article, further inquiries can be directed to the corresponding author.

Conflicts of Interest: The authors declare no conflicts of interest.

Abbreviations

The following abbreviations are used in this manuscript:

HV	high-voltage
FET	field-effect transistor
SiC	silicon carbide
MOSFET	metal-oxide-semiconductor field-effect transistor
PCB	printed circuit board
IGBT	insulated-gate bipolar transistor
SOA	safe operating area

References

1. Zhuge, Y.; Liang, J.; Fu, M.; Long, T.; Wang, H. Comprehensive overview of power electronics intensive solutions for high-voltage pulse generators. *IEEE Open J. Power Electron.* **2023**, *5*, 1–20. [CrossRef]
2. Abadi, M.R.Q.R.; Marzebali, M.R.; Abolghasemi, V.; Anisi, M.H. High-voltage pulse generators for electroporation application: A systematic review. *IEEE Access* **2022**, *10*, 64933–64951. [CrossRef]
3. Withanage, R.; Shammas, N. Series connection of insulated gate bipolar transistors (IGBTs). *IEEE Trans. Power Electron.* **2012**, *27*, 2204–2212. [CrossRef]
4. Frivaldsky, M.; Zelnik, R.; Spanik, P. Design proposals for high-voltage stacked configuration GaN module. *Electronics* **2024**, *13*, 3198. [CrossRef]
5. Zhao, W.; Ghafoor, S.; Lagerweij, G.W.; Rietveld, G.; Vaessen, P.; Niasar, M.G. Comprehensive investigation of promising techniques to enhance the voltage sharing among SiC MOSFET strings, supported by experimental and simulation validations. *Electronics* **2024**, *13*, 1481. [CrossRef]
6. Langpoklakpam, C.; Liu, A.-C.; Chu, K.-H.; Hsu, L.-H.; Lee, W.-C.; Chen, S.-C.; Sun, C.-W.; Shih, M.-H.; Lee, K.-Y.; Kuo, H.-C. Review of silicon carbide processing for power MOSFET. *Crystals* **2022**, *12*, 245. [CrossRef]
7. Chen, Z.; Huang, A.Q. Extreme high efficiency enabled by silicon carbide (SiC) power devices. *Mat. Sci. Semicon Proc.* **2024**, *172*, 108052. [CrossRef]
8. Available online: https://www.wolfspeed.com/products/power/sic-mosfets/1700v-silicon-carbide-mosfets/c2m1000170d/ (accessed on 11 February 2025).
9. Available online: https://www.st.com/en/power-transistors/stp4n150.html (accessed on 11 February 2025).
10. Available online: https://iganpower.com/ganhemts (accessed on 11 February 2025).

11. Available online: https://www.infineon.com/cms/en/product/power/mosfet/silicon-carbide/discretes/imyh200r100m1h/ (accessed on 11 February 2025).
12. Available online: https://navitassemi.com/genesic-mosfets-products/ (accessed on 11 February 2025).
13. Available online: https://www.littelfuse.com/products/power-semiconductors-control-ics/mosfets-si-sic/n-channel-standard/very-high-voltage (accessed on 11 February 2025).
14. Available online: https://www.qorvo.com/design-hub/design-tools/interactive/qspice (accessed on 11 February 2025).
15. Bolotnikov, A.; Losee, P.; Permuy, A.; Dunne, G.; Kennerly, S.; Rowden, B.; Stevanovic, L. Overview of 1.2kV–2.2kV SiC MOSFETs targeted for industrial power conversion applications. In Proceedings of the 2015 IEEE Applied Power Electronics Conference and Exposition (APEC), Charlotte, NC, USA, 15–19 March 2015; pp. 2445–2452. [CrossRef]
16. Bemius, M.T.; Chutjian, A. High-voltage, full-floating 10-MHz square-wave generator with phase control. *Rev. Sci. Instrum.* **1989**, *60*, 779–782. [CrossRef]
17. Baker, R.J.; Johnson, B.P. Stacking power MOSFETs for use in high-speed instrumentation. *Rev. Sci. Instrum.* **1992**, *63*, 5799–5801. [CrossRef]
18. Available online: https://www.behlke.com/pdf/general.pdf (accessed on 21 February 2025).
19. Available online: https://www.paramerus.com/pages/hvpg.php (accessed on 21 February 2025).
20. Abbate, C.; Busatto, G.; Iannuzzo, F. High-voltage, high-performance switch using series-connected IGBTs. *IEEE Trans. Power Electron.* **2010**, *25*, 2450–2459. [CrossRef]
21. Rutten, T.P.; Wild, N.; Veitch, P.J. Fast rise time, long pulse width, kilohertz repetition rate Q-switch driver. *Rev. Sci. Instrum.* **2007**, *78*, 073108. [CrossRef] [PubMed]
22. Son, M.; Lee, T.; Kwon, S.; Cho, Y. Impact of snubber parameters on voltage sharing in series-connected insulated gate bipolar transistors. *J. Pow. Elect.* **2020**, *20*, 1002–1014. [CrossRef]
23. Alves, F.S.; Lefranc, P.; Jeannin, P.-O.; Sarrazin, B. Advanced voltage balancing techniques for series-connected SiC-MOSFET devices: A comprehensive survey. *Power Elect. Dev. Compon.* **2024**, *7*, 100055. [CrossRef]
24. Wang, R.; Jørgensen, A.B.; Liu, W.; Zhao, H.; Yan, Z.; Munk-Nielsen, S. Voltage balancing of series-connected SiC mosfets with adaptive-impedance self-powered gate drivers. *IEEE Trans. Indust. Electron.* **2023**, *70*, 11401–11411. [CrossRef]
25. Despotuli, A.L.; Kazmiruk, V.V.; Despotuli, A.A.; Andreeva, A.V. Generator of High-Voltage Rectangular Pulses. Patent RU № 2 819 031. Available online: https://patenton.ru/patent/RU2819031C1 (accessed on 11 February 2025).
26. Despotuli, A.L.; Kazmiruk, V.V.; Despotuli, A.A.; Andreeva, A.V. A generator of rectangular pulses based on a serial connection of mosfet with U_{max} = 4500 V. *Instrum. Exp. Technol.* **2024**, *67*, 54–61. [CrossRef]
27. Persson, E.; Di Domenico, F. Optimizing PCB layout for HV CoolGaN power transistors. AN_2306_PL52_2308_151809. Available online: https://www.infineon.com/dgdl/Infineon-CoolGaN_High_voltage_power_transistors_for_PCB_layout_optimization-ApplicationNotes-v01_00-EN.pdf?fileId=8ac78c8c8929aa4d0189ba64ffb262a2&da=t (accessed on 11 February 2025).
28. Available online: https://files.resources.altium.com/sites/default/files/2022-09/file_665[1].pdf (accessed on 11 February 2025).
29. Okuda, T.; Nishimura, Y.; Nishioka, K.; Kishimoto, S.; Kikuchi, Y.; Nakamura, T. A 5-kV pulse generator with a 100-kV/µs slew rate based on series-connected 1700-V SiC MOSFETs for electrical insulation tests. *Rev. Sci. Instrum.* **2021**, *92*, 114705. [CrossRef] [PubMed]
30. Achour, Y.; Starzynski, J.; Jakubiuk, K. New architecture of solid-state high-voltage pulse generators. *Energies* **2022**, *15*, 4823. [CrossRef]
31. Novickij, V.; Rembiałkowska, N.; Szlasa, W.; Kulbacka, J. Does the shape of the electric pulse matter in electroporation? *Front. Oncol.* **2022**, *12*, 958128. [CrossRef] [PubMed]
32. Knuteson, B. A quantitative measure of experimental scientific merit. Available online: https://arxiv.org/abs/0712.3572 (accessed on 11 February 2025).
33. Fortunato, S.; Bergstrom, C.T.; Börner, K.; Evans, J.A.; Helbing, D.; Milojević, S.; Petersen, A.M.; Radicchi, F.; Sinatra, R.; Uzzi, B.; et al. Science of science. *Science* **2018**, *359*, aao0185. [CrossRef] [PubMed]

energies

MDPI

Article

Research on Missing Data Estimation Method for UPFC Submodules Based on Bayesian Multiple Imputation and Support Vector Machines

Xiaoming Yu [1], Jun Wang [1], Ke Zhang [1], Zhijun Chen [1], Ming Tong [1], Sibo Sun [2], Jiapeng Shen [2,*], Li Zhang [2] and Chuyang Wang [2]

[1] State Grid Suzhou Power Supply Company, Suzhou 215004, China; jstzyxm@163.com (X.Y.); wangjun@163.com (J.W.); zhangke@163.com (K.Z.); laugher_007@sina.com (Z.C.); colla_t@163.com (M.T.)

[2] College of Electrical and Power Engineering, Hohai University, Nanjing 211106, China; ss13898394826@163.com (S.S.); zhanglinuaa@hhu.edu.cn (L.Z.); wangchuyang@hhu.edu.cn (C.W.)

* Correspondence: 210206030004@hhu.edu.cn

Abstract: With the increasing complexity of power systems, the monitoring data of UPFC submodules suffers from high missing rates due to sensor failures and environmental interference, significantly limiting equipment condition assessment and fault warning capabilities. To overcome the computational complexity, poor real-time performance, and limited generalization of existing methods like GRU-GAN and SOM-LSTM, this study proposes a hybrid framework combining Bayesian multiple imputation with a Support Vector Machine (SVM) for data repair. The framework first employs an adaptive Kalman filter to denoise raw data and remove outliers, followed by Bayesian multiple imputation that constructs posterior distributions using normal linear correlations between historical and operational data, generating optimized imputed values through arithmetic averaging. A kernel-based SVM with RBF and soft margin optimization is then applied for nonlinear calibration to enhance robustness and consistency in high-dimensional scenarios. Experimental validation focusing on capacitor voltage, current, and temperature parameters of UPFC submodules under a 50% missing data scenario demonstrates that the proposed method achieves an 18.7% average error reduction and approximately 30% computational efficiency improvement compared to single imputation and traditional multiple imputation approaches, significantly outperforming neural network models. This study confirms the effectiveness of integrating Bayesian statistics with machine learning for power data restoration, providing a high-precision and low-complexity solution for equipment condition monitoring in complex operational environments. Future research will explore dynamic weight optimization and extend the framework to multi-source heterogeneous data applications.

Keywords: unified power flow controller (UPFC); Bayesian multiple imputation; support vector machine; data restoration

1. Introduction

With the ongoing expansion of power system scales and increasing penetration of renewable energy, modern grids face growing demands for real-time and precise power flow control [1,2]. As a core device in FACTS, the UPFC enhances power transmission efficiency and system stability by dynamically regulating transmission line impedance, voltage magnitude, and phase angle differences [3,4]. Driven by MMC technology, MMC-UPFC has emerged as critical infrastructure in ultra-high-voltage transmission projects due to its modular design, low harmonic output, and high reliability [5–7]. However, MMC-UPFC submodule monitoring data face significant challenges in complex operational environments. Firstly, the microsecond-level switching of hundreds of submodules imposes stringent requirements on data acquisition accuracy. Secondly, high-frequency electromagnetic interference and harmonic pollution degrade the signal-to-noise ratio of monitoring data, with typical reductions exceeding 40% under operational conditions [8–10]. Additionally, sensor failures and environmental disturbances result in missing rates of 15–20% for critical parameters such as capacitor voltage, current, and temperature, leading to "black module" phenomena that severely hinder equipment condition assessment and fault warning capabilities [11,12].

Current data restoration methods exhibit significant limitations in addressing these challenges. While conventional statistical imputation techniques (e.g., linear interpolation, mean filling) offer computational simplicity, they neglect the spatiotemporal correlations and nonlinear characteristics of power data, resulting in large dynamic restoration errors during transient conditions [13]. Although neural network-based approaches such as GRU-GAN and LSTM can capture complex spatiotemporal dependencies, their training relies on complete datasets, and their high model complexity (e.g., GRU-GAN requiring simultaneous optimization of generators and discriminators) hampers real-time processing on embedded platforms [14–16]. Furthermore, traditional multiple imputation techniques reduce uncertainty through multiple imputed values but fail to account for the non-Gaussian distribution characteristics of power data, leading to posterior estimation biases. Additionally, existing methods primarily focus on single-time-scale restoration, lacking adaptability to the multiscale dynamics of UPFC submodules: microsecond-level capacitor voltage fluctuations demand high-precision imputation for real-time control, whereas hour-level temperature trend analysis requires global consistency, posing dual challenges to algorithm robustness and computational efficiency [17,18].

To address the aforementioned issues, this study proposes a hierarchical repair framework based on Bayesian multiple imputation and a Support Vector Machine, which aims to construct an adaptive imputation matrix by establishing a dynamic probabilistic model to analyze the multidimensional operational parameter distribution characteristics of UPFC equipment. While maintaining temporal correlation of data, the framework effectively overcomes the error accumulation problem inherent in traditional methods for power electronic device data repair. Innovatively employing a feature space mapping mechanism, the framework incorporates both the capacitor voltage fluctuation characteristics of UPFC submodules and the switching states of power electronic devices as imputation constraints. This approach achieves enhanced reconstruction accuracy for missing values while maintaining low computational complexity through dimensionality reduction operations, ultimately providing an engineering-practical data preprocessing solution for reliability analysis in flexible AC transmission systems., with core contributions including adaptive data preprocessing through modified Kalman filtering for dynamically adjusting noise covariance matrices to enhance data quality in high-noise scenarios; implementation of multiscale Bayesian multiple imputation that innovatively incorporates discrete-time equations derived from UPFC submodule state–space modeling into posterior distributions and missing value estimation processes, integrating historical data with real-time observations to generate multiple imputation sets while achieving coordinated imputation for transient and sustained missing data through posterior optimization; and nonlinear SVM calibration employing RBF kernels to map high-dimensional feature spaces, utilizing soft-margin optimization to eliminate outliers in imputation results and reinforce physical consistency in restored data. However, it should be noted that the Bayesian multiple imputation method still heavily relies on the completeness and representativeness of the dataset. Specifically, the imputation model constructed in this study imposes strict accuracy constraints on the measured parameters of the MMC-UPFC topology (including submodule capacitance values and arm inductance), where parameter deviations exceeding ±2% will lead to a significant expansion of the confidence intervals in the imputation results.

The first section elucidates the key feature extraction of UPFC submodules, initiating from submodule operating modes by adopting average switching functions to solve MMC submodule characteristic equations and employing the fourth-order Runge–Kutta method for discretization. The second section introduces fundamental principles of Bayesian multiple imputation and the missing value imputation process, incorporating UPFC submodule characteristics. The third section systematically elaborates the Support Vector Machine (SVM)-based calibration methodology for UPFC missing data, which performs secondary calibration and optimization on Bayesian multiple imputation results, with comparative experimental validation demonstrating significant enhancements in high-dimensional data feature resolution and nonlinear relationship modeling accuracy compared to conventional imputation frameworks. The fourth section addresses missing value imputation for UPFC from both short-term and long-term scales to validate the algorithm's effectiveness.

2. UPFC Submodule Key Data Feature Extraction

2.1. Submodule Modal and Waveform Analysis

Figure 1 illustrates the schematic of a three-phase MMC-UPFC. Each phase arm comprises N SMs adopting a half-bridge configuration, where S1 and S2 represent the switching devices, and C_{SM} denotes the submodule capacitance. L_o represents the arm filter inductor.

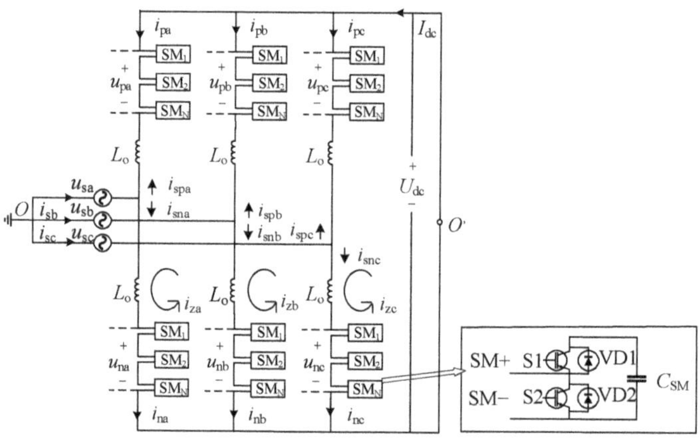

Figure 1. Three-phase MMC-UPFC topological schematic diagram.

Figure 2 illustrates the operating state diagram of SMs in an MMC-UPFC. SM operational states are categorized into three types: inserted, bypassed, and blocked. In the inserted state, the operation is divided into Mode 1 and Mode 2 based on whether the arm current I_{pj} is positive ($i_{pj} > 0$) or negative. In the bypassed state, switching device T1 remains off while T2 is on, with its operation (Modes 3 and 4) exhibiting an inverse switching configuration compared to the inserted state. For the blocked state, both T1 and T2 are turned off, and SM behavior is entirely dictated by the direction of the arm current, resulting in Modes 5 and 6. The current flow directions for all modes are explicitly indicated in Figure 2.

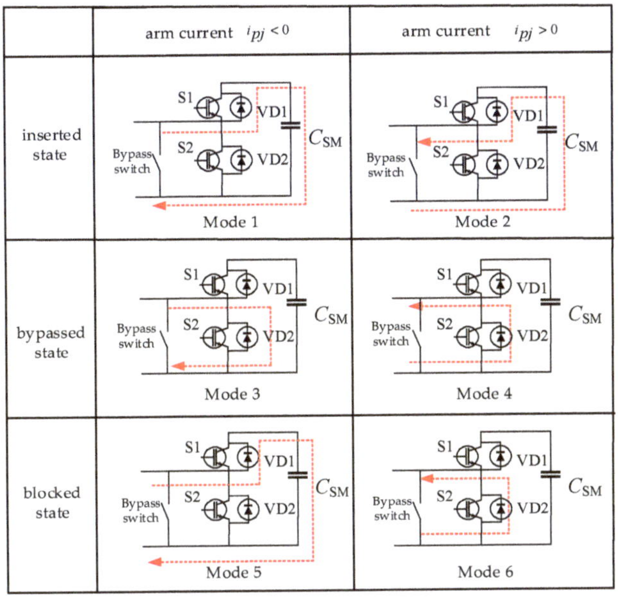

Figure 2. MMC submodule operating modes.

During operation, the MMC consistently operates in three switching states. For the non-blocked state, the switching functions of the *i*-th submodule in the upper and lower arms (denoted as *p* and *n*) of phase *j* (*j* = *a*, *b*, *c*) are defined as follows:

$$S_{SMi,jp(n)}(t) = \begin{cases} 1, S = 1 \\ 0, S = 0 \end{cases} \quad (1)$$

In the equation, *S* represents the submodule state, where *S* = 1 indicates an inserted submodule and *S* = 0 otherwise. By combining the relational expressions of all submodules in phase *j*, performing Fourier decomposition on the switching functions of the upper and lower arms while neglecting higher-order harmonic terms, the fundamental-frequency average switching function expressions for the MMC upper and lower arms can be derived as

$$S(t) = \begin{cases} \frac{1-M\sin(\omega_1 t)}{2} \text{ Upper arm} \\ \frac{1+M\sin(\omega_1 t)}{2} \text{ Lower arm} \end{cases} \quad (2)$$

In the equation, *M* represents the modulation ratio.

Under CPS-SPWM strategy, the KVL equations for the upper and lower arms in a single-phase unit are formulated as follows:

$$L_o\frac{d(\frac{1}{2}i_{sj})}{dt} + L_o\frac{di_{zj}}{dt} = u_{oo'} + \frac{U_{dc}}{2} - u_{sj} - u_{pj} \quad (3)$$

$$L_o\frac{d(\frac{1}{2}i_{sj})}{dt} - L_o\frac{di_{zj}}{dt} = u_{oo'} - \frac{U_{dc}}{2} - u_{sj} + u_{nj} \quad (4)$$

$$I_{sj} = I_{nj} - I_{pj} \quad (5)$$

$$I_{zj} = \frac{P}{3U_{dc}} = \frac{1}{2}(I_{nj} + I_{pj}) \quad (6)$$

In the equation, the reference point is set as point O, the grid-side power supply reference as point O′, with $U_{OO'}$ representing the voltage between these two points, and L_O denoting the bridge arm inductance value; I_{pj} and I_{nj} represent the current amplitudes of the upper and lower arms, respectively; I_{sj} denotes the amplitude of the grid-side AC current; i_{zj} is the inter-phase circulating current; I_{zj} stands for its amplitude; and U_{dc} corresponds to the DC capacitor voltage.

2.2. Key Data Feature Extraction

Submodule critical data primarily consist of the submodule capacitor voltage and arm input current, as shown in Figure 3. Through analysis of the short-timescale dynamic behavior of UPFC submodules, the mathematical representations of these critical data characteristics can be derived through the integration of switching functions with the KVL equations governing the upper and lower converter arms.

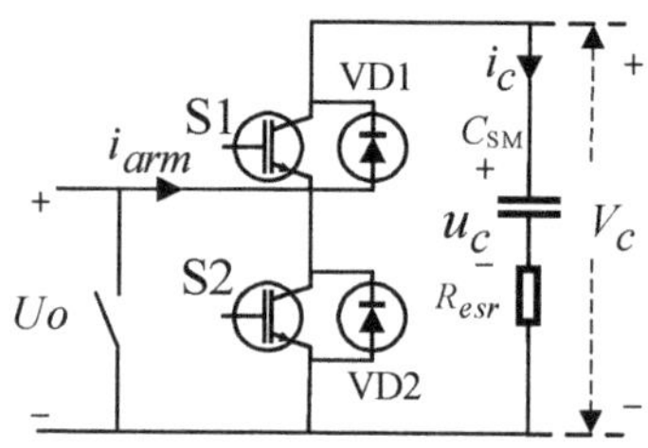

Figure 3. Detailed submodule topology.

The submodule current i_c, submodule voltage V_c, and submodule capacitor voltage u_c are defined as

$$\begin{cases} i_c = S(t) \cdot i_{arm}(t) \\ i_{arm}(t) \approx \frac{I_{dc}}{3} - \frac{I_S}{2}\cos(\omega_1 t + \varphi_1) - \frac{MI_S}{4}\cos(\omega_1 t + \varphi_2) \\ V_c = \frac{U_{dc}}{N} \\ u_c = V_c - i_c R_{esr} = \frac{U_{dc}}{N} + \frac{1}{C}\int_{t=0}^{\infty} i_c dt - i_c R_{esr} \end{cases} \tag{7}$$

In the equation, i_{arm} denotes the arm current; $I_{dc}/3$ represents the DC-side input current; I_S and φ_1 correspond to the magnitude and phase angle of the MMC-UPFC AC-side current; $MIs/4$ and φ_2 specify the amplitude and phase angle of the second-harmonic circulation component; R_{esr} denotes the capacitor's equivalent series resistance; and C_{SM} denotes the DC-side capacitance value.

Taking the upper arm as an example, the differential expressions governing the submodule capacitor voltage u_c and submodule current i_c are derived from the above relationships as

$$\begin{cases} \frac{du_c}{dt} = \frac{1}{C_{SM}} i_c - R_{esr}\frac{di_c}{dt} \\ \frac{di_c}{dt} = S(t)\frac{di_{arm}(t)}{dt} + i_{arm}(t)\frac{dS(t)}{dt} \end{cases} \tag{8}$$

$$\frac{du_c}{dt} = \frac{1}{C_{SM}} i_c - R_{esr}\left(S(t)\frac{di_{arm}(t)}{dt} + i_{arm}(t)\frac{dS(t)}{dt}\right) \tag{9}$$

The mathematical model of the MMC-UPFC submodule can be discretized via the fourth-order Runge–Kutta method, and the above equation can be expressed as

$$f_1(t, u_c) = \frac{du_c}{dt} = \frac{1}{C_{SM}} i_c - R_{esr}\left(S(t)\frac{di_{arm}(t)}{dt} + i_{arm}(t)\frac{dS(t)}{dt}\right) \tag{10}$$

The state variable update equations for the next sampling period are formulated as follows:

$$u_{c,n+1} = u_{c,n} + \frac{h}{6}(k_{a1} + 2k_{a2} + 2k_{a3} + k_{a4}) \tag{11}$$

In the equation, h denotes the computational step size and K_{a1}~K_{a4} represent the operational coefficients, with their specific formulas expressed as follows:

$$\begin{cases} k_{a1} = f_1(t, u_c) \\ k_{a2} = f_1(t + \frac{h}{2}, u_c + \frac{h}{2}k_{a1}) \\ k_{a3} = f_1(t + \frac{h}{2}, u_c + \frac{h}{2}k_{a2}) \\ k_{a4} = f_1(t + h, u_c + hk_{a3}) \end{cases} \tag{12}$$

The state–space equations can be formulated as

$$u_{c,n+1} = f(C_{SM}, R_{esr}) \cdot u_{c,n} \tag{13}$$

3. Submodule Data Imputation Based on Multiple Bayesian Imputation

3.1. Data Acquisition and Preprocessing

Prior to data imputation, preprocessing of UPFC submodule detection data is implemented through Kalman filtering to standardize data formats.

This methodology incorporates cascaded filtering architecture to eliminate both measurement-specific noise from submodule sensors and systemic interference inherent in power electronic systems. Following noise suppression, dynamic threshold determination is executed through historical data averaging, where data points exceeding preset multiples of the mean value are classified as outliers for systematic removal. Multi-sensor redundancy validation ensures measurement reliability by performing cross-correlation analysis on concurrent sampling datasets from identical measurement nodes [19]. The formula of the adaptive Kalman filter is as follows:

$$\begin{cases} \hat{x}_{\mathrm{k}} = \hat{x}_{\mathrm{k|k-1}} + G_{\mathrm{k}}(z_{\mathrm{k}} - H_{\mathrm{k}}\hat{x}_{\mathrm{k|k-1}}) \\ \hat{x}_{\mathrm{k|k-1}} = F_{\mathrm{k-1}}\hat{x}_{\mathrm{k-1}} \\ G_{\mathrm{k}} = P_{\mathrm{k|k-1}}H_{\mathrm{k}}^{\mathrm{T}}(H_{\mathrm{k}}P_{\mathrm{k|k-1}}H_{\mathrm{k}}^{\mathrm{T}}/\beta_{\mathrm{k}} + R_{\mathrm{k}})^{-1}/\beta_{\mathrm{k}} \\ P_{\mathrm{k|k-1}} = F_{\mathrm{k-1}}P_{\mathrm{k-1}}F_{\mathrm{k-1}}^{\mathrm{T}} + Q_{\mathrm{k-1}} \\ P_{\mathrm{k}} = (I - G_{\mathrm{k}}H_{\mathrm{k}})P_{\mathrm{k|k-1}}/\beta_{\mathrm{k}} \end{cases} \tag{14}$$

The formula defines the following key parameters of adaptive Kalman filtering: $\hat{x}_k$ denotes the estimated state vector at time step k, $\hat{x}_{k-1}$ represents the previous estimated state vector, $\hat{x}_{k|k-1}$ indicates the one-step predicted state vector, G_{k} is the adaptive Kalman gain matrix, z_{k} corresponds to the measurement vector, H_{k} stands for the measurement matrix, $F_{\mathrm{k-1}}$ denotes the state transition matrix at step $\mathrm{k} - 1$, $P_{\mathrm{k|k-1}}$ defines the one-step predicted state error covariance matrix, P_{k} represents the estimated state error covariance matrix, β_{k} serves as the adaptive factor, and R_{k} and $Q_{\mathrm{k-1}}$, respectively, characterize the measurement noise covariance matrix and process noise covariance matrix, with I being the identity matrix.

The parameter values exceeding the preset multiple of standard deviations from the mean are defined as outliers, where the selection of parameter data points is dynamically adjusted according to the temporal characteristics and variation patterns of the parameter's operational time scale.

3.2. Bayesian Multiple Imputation

3.2.1. Definition of Missing Values

Following data preprocessing, detection is performed on both operational and historical datasets to identify and locate missing data points, enabling data supplementation and refinement to establish a repaired operational dataset [20].

The UPFC submodule monitoring data comprises missing variables Y and non-missing variables X, including the capacitor voltage, submodule positive terminal voltage, and submodule temperature values. After removing the outliers, historical and operational data are partitioned into equivalent periodic intervals based on state parameters. Each interval is divided into n data points, with the interval length set to correspond to one or multiple cycle durations of the state parameter, adjustable according to data completeness. All complete datasets partitioned under these parameter-specific periodic intervals are aggregated into set X, formatted as an $n \times q$ matrix where q denotes the number of complete datasets. Set X represents the non-missing variables in the UPFC submodule monitoring data.

The UPFC submodule monitoring data are represented by missing variable Y, which corresponds to an arbitrarily selected incomplete dataset partitioned according to equivalent periodic intervals of the state parameter. Dataset Y constitutes either a historical dataset with missing entries from a specific time period or an incomplete current operational dataset, serving to characterize the missing variables in the UPFC submodule monitoring data.

The variables Y and X possess a linear relationship characterized by a normal linear correlation between their parameters, expressed as follows:

$$Y \sim N\left(X\beta, \sigma^2\right) \tag{15}$$

In the equation, β denotes the regression coefficient, which is a q-dimensional vector, and σ^2 represents the residual variance.

In dataset Y, the n_1 non-missing observed values are denoted as Y_1, and the corresponding portion of X associated with these observations is labeled X_1. The n_0 artificially created missing values in Y are recorded as Y_{m}, with their corresponding X portion designated as X_{m}.

3.2.2. Construction of the Posterior Distribution

Based on the non-missing data portion, the parameters characterizing the relationship between Y and X are estimated, and the posterior distribution of these parameters is constructed. Subsequently, for each missing data point, samples are drawn from this posterior distribution to obtain imputed estimates for the missing values.

In dataset Y, the n_1 non-missing observed values are denoted as Y_1, with their corresponding portion in X labeled X_1; and the n_0 missing values in Y are recorded as Y_m, with their corresponding X portion is designated X_m. Based on the non-missing data subset, estimation is performed using the following formula [21]:

$$\begin{aligned} \hat{\beta}_1 &= (X_1^T X_1)^{-1}(X_1^T Y_1) \\ \hat{\sigma}_1^2 &= (Y_1 - X_1\hat{\beta}_1)^T (Y_1 - X_1\hat{\beta}_1)/(n_1 - q) \end{aligned} \tag{16}$$

In the equation, $\hat{\beta}_1$ denotes the estimated regression coefficients of the model, and $\hat{\sigma}_1^2$ represents the estimated residuals.

Within the Bayesian framework, the posterior distributions of β and the residuals can be constructed based on OLS estimates. The posterior distribution of the estimated residuals α^2 chi-square distribution properties are as follows:

$$\frac{\hat{\sigma}_1^2}{\sigma^2}(n-q) \sim \chi^2(n-q) \tag{17}$$

In the equation, n represents the number of observed values of state parameters within an equidistant cycle.

The posterior distribution of β depends on the estimated residuals α^2, following a normal distribution.

3.2.3. Imputation of Missing Values

Missing values are estimated by sampling from the posterior distribution and are combined with the imputed dataset generated by the following imputation model:

$$Y_* = X_m\beta_* + Z_1\sigma_* \tag{18}$$

In the equation, Y_* is a set of imputed values; σ* is a random sample drawn from the posterior distribution; β^* is a random sample drawn from the conditional posterior distribution when $\sigma = \sigma*$; and Z_1 consists of n_0 independent random standard normal variates, expressed as

$$\sigma_*^2 = \sigma_1^2(n_1 - q)/g \tag{19}$$

$$\beta_* = \hat{\beta}_1 + \sigma_*\left(\left(X_1^T X_1\right)^{-1}\right)^{0.5} Z_2 \tag{20}$$

In the equation, g is generated by a random draw from χ^2 $(n_1 - q)$, and Z_2 consists of q independent random standard normal variates.

Considering the critical data characteristics of UPFC submodules, the revised imputation model is formulated as

$$Y_*{}' = Y_* + \lambda_0 Y_{comp} \tag{21}$$

In the equation, λ_0 is the proportionality coefficient derived from the imputation model based on the critical data characteristics of the UPFC submodules; and Y_{comp} represents the corrected results according to the key physical quantities of the UPFC submodules, where each $y_{comp,i} \in Y_{comp}$ is specifically defined as

$$y_{comp,i} = m\gamma_1 f(C_{sm}, R_{esr})y(k+1) + \gamma_2 f(C_{sm}, R_{esr})y(k-1) i = 1, 2 \ldots, m \tag{22}$$

In the equation, k represents the position of the i-th missing data in Y_m within the original dataset Y. $y(k + 1)$ and $y(k - 1)$ denote the data at positions $k + 1$ and $k - 1$ in Y, respectively; and γ_1 and γ_2 are determined by the time scale and the information status of data at the missing position, where the formula is provided by the state of Equation (13) in Section 2.2. The determination method is illustrated in the Figure 4 below.

The repaired dataset Y' is obtained by calculating the arithmetic mean of all imputed datasets.

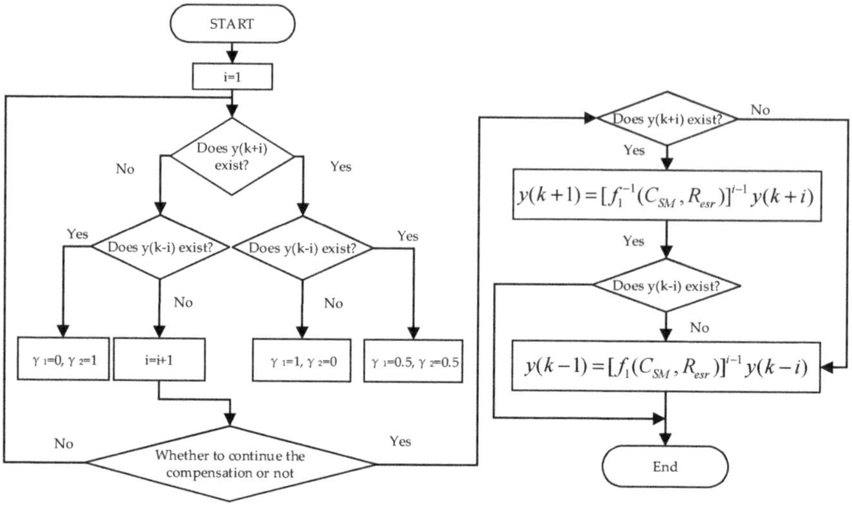

Figure 4. Missing value imputation process framework.

4. SVM-Based Evaluation of Imputed Data

In the formulation, the SVM is a supervised learning algorithm grounded in statistical learning theory, primarily employed for data classification and regression tasks. Its core principle involves locating an optimal hyperplane to separate data of different classes while maximizing the margin, defined as the distance between the hyperplane and the nearest data points [22].

In power system analysis, preprocessed and multiply imputed UPFC data are typically nonlinearly separable. To enable SVMs to process such data, the dataset must be projected into a higher-dimensional space where linear separation becomes feasible. This necessitates the application of kernel techniques in SVMs, whose fundamental principle involves implicitly mapping data to high-dimensional spaces through kernel functions without explicitly computing coordinates in the transformed space. The kernel function essentially computes the inner product between two data points in the high-dimensional space, expressed as

$$K(xi,xj) = \phi(xi) \cdot \phi(xj) \tag{23}$$

In the equation, $\phi(\bullet)$ represents the mapping function that maps data from the original space into a high-dimensional space.

4.1. SVM-Based Data Calibration

The repaired operational dataset is denoted as Y'. For a given set of data points (x_i, y_i) where $i = 1, 2, \ldots, l$, with $x_i \in Y'$ and $y_i \in \{-1, 1\}$, an optimal hyperplane can be identified to separate the samples when linear separability is achieved, mathematically represented as

$$w^T x + b = 0 \tag{24}$$

In the equation, w and b represent the weight vector and the optimal hyperplane offset, respectively. The optimal hyperplane is shown in Figure 5 below.

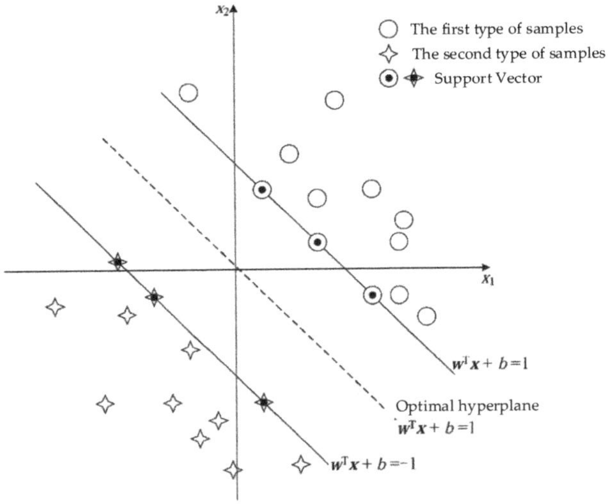

Figure 5. Support vector machine optimal hyperplane.

The SVM method utilizes soft margin optimization and kernel tricks to map samples from the original space into a high-dimensional feature space. Within this high-dimensional space, the SVM identifies a maximum-margin linear hyperplane for data classification. This linear hyperplane is formulated by solving the following optimization problem:

$$\begin{cases} \min\limits_{w,b} & \frac{1}{2}\|w\|^2 + C\sum\limits_{i=1}^{m} \xi_i \\ \text{s.t.} & y_i[w^T \cdot \varphi(x_i) + b] \geq 1 - \xi_i, \xi_i \geq 0, i = 1, \ldots, m \end{cases} \tag{25}$$

In the above equation, $C > 0$ denotes the penalty factor, ξ_i represents the slack variable, and $\varphi(x_i)$ corresponds to the feature vector of x_i mapped into the high-dimensional space. In the calculation process of the above equation, $\varphi(x_i)^T\varphi(x_i)$ needs to be solved. It is solved by using the kernel function [23]. In the Support Vector Machine (SVM) method, there are four types of kernel functions, which are mathematically defined as follows:

$$\begin{cases} \text{linear} : K(x_i, x_j) = {x_i}^T x \\ \text{polynomial} : K(x_i, x_j) = (\gamma {x_i}^T x + r)^d, \gamma > 0 \\ \text{RBF} : K(x_i, x_j) = \exp(-\gamma\|x_i - x\|^2), \gamma > 0 \\ \text{sigmoid} : K(x_i, x_j) = \tanh(\gamma {x_i}^T x + r) \end{cases} \tag{26}$$

In the above equation, r, γ, and d are all kernel parameters.

4.2. The Advantages of Applying Support Vector Machines (SVMs) to UPFC Submodule Imputation

For sensor data such as UPFC submodules, which typically consist of multi-parameter and multi-dimensional numerical values, SVMs implicitly map such data into high-dimensional spaces via the kernel trick, making them suitable for handling high-dimensional features. Even when the feature dimensionality far exceeds the sample size, SVMs can maintain generalization performance by maximizing the margin, thereby avoiding the "curse of dimensionality".

UPFC data involve complex nonlinear relationships such as coupling between voltage and power. SVMs can effectively capture these relationships using nonlinear kernel functions like the Gaussian kernel, thereby further improving imputation accuracy. If redundant features exist in high-dimensional data, SVM feature weights (e.g., linear kernel coefficients) can be utilized to screen critical variables, enhancing imputation efficiency.

5. Case Analysis

5.1. Data Source

5.1.1. Short-Time-Scale Data Sources

Simulation data were acquired through PLECS-based modeling of the UPFC, with the corresponding simulation model configuration depicted in the following Figure 6 and critical operational parameters specified in the accompanying Table 1:

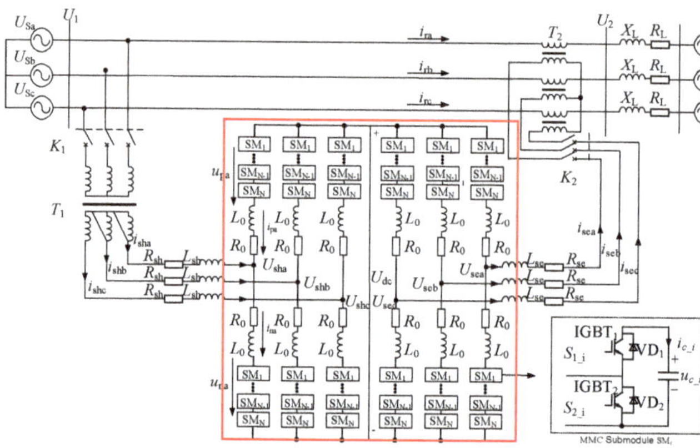

Figure 6. UPFC operational topology for simulation.

Table 1. Key parameters.

Parameter Name	Value
Number of Submodules (N)	20
AC-side Voltage Frequency (fg)	50 Hz
AC-side Line Voltage Amplitude (*U*gm)	500 kV
DC Bus Total Reference Voltage (*U*dc)	180 kV
DC Capacitor Initial Voltage per Submodule (*U*sm)	Udc/N (kV)
Submodule Capacitance (Csm)	11×10^{-3} F
DC Bus Capacitance (Carm)	100×10^{-6} F
Filter Inductance (*L*arm)	36×10^{-3} H
Filter Winding Resistance (*R*arm)	0.1 Ω
Equivalent Inductance (*L*g)	1.00×10^{-3} H
Equivalent Resistance (*R*g)	0.1 Ω
Carrier Ratio (FR)	2.4
Triangular Carrier Frequency (*f*sw)	Fg × FR Hz
Line Inductance (*L*n)	0.001 H
Line Resistance (*R*n)	1 Ω
Line Capacitance (C_L)	1.00×10^{-5}

The simulation parameters are configured as follows: The UPFC series-side capacity is 250 × 2 MVA, with the shunt-side capacity set at 250 MVA and an AC bus voltage regulation range spanning 485 kV to 535 kV. The maximum operating voltage at the converter valve-side reaches 105 kV for the series unit and 94 kV for the shunt unit, while the DC-side maintains a rated voltage of 90 kV with a maximum operational voltage of 92.5 kV. This configuration strictly adheres to IEC 61803 standards for power electronic converter testing.

5.1.2. Long-Term Time-Scale Data Source

The long-term time-scale data source originates from real UPFC phasor measurement data, where the phasor measurement data are selected from 1 October 2023, as shown in Figure 7.

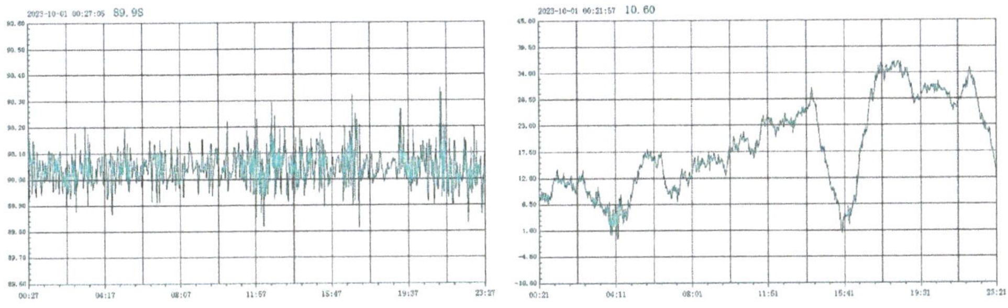

(**a**) (**b**)

Figure 7. (**a**) The telemetry voltage measurements, and (**b**) the telemetry current measurements recorded on 1 October 2023.

5.2. Process Framework

Figure 8 illustrates the framework of the data restoration process. Time scales are distinguished based on data sampling intervals. When short-term time-scale data are identified, missing data restoration is executed by integrating key operational characteristics of UPFC submodules with numerical analysis methods. For long-term time-scale missing data scenarios where UPFC circuit intrinsic characteristics become non-manifested, data imputation is exclusively performed through numerical analysis methods.

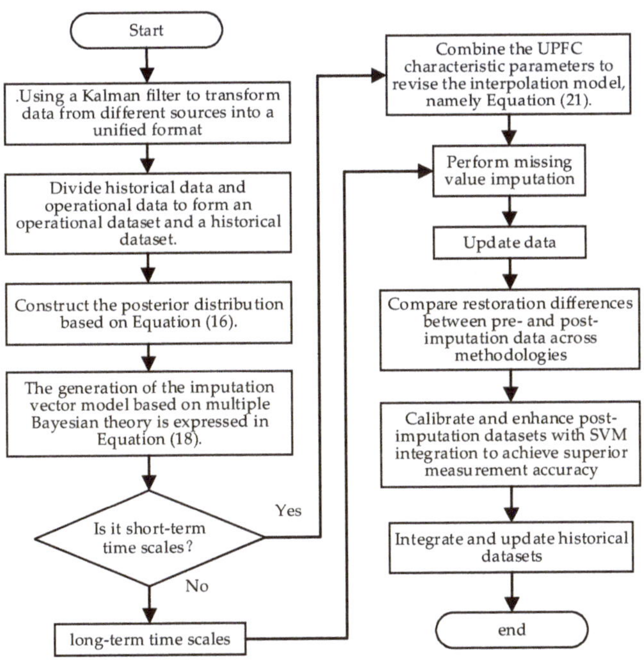

Figure 8. Data restoration process framework.

5.3. Imputation Data Accuracy Validation Metrics

Following data imputation, a comprehensive accuracy assessment between the imputed and original datasets was performed using four statistical parameters—Mean Squared Error (MSE), Mean Absolute Error (MAE), Pearson correlation coefficient (r), and Coefficient of determination (R^2)—to evaluate the precision of the imputation results.

The MSE quantifies the average squared deviation between imputed and true values, demonstrating sensitivity to significant errors with its optimal value approaching zero. The MAE measures the absolute average magnitude of imputation errors, providing intuitive error magnitude interpretation, where values below 15% of the standard deviation indicate satisfactory performance. The r evaluates linear correlation strength between imputed and true values, with absolute values exceeding 0.8 denoting strong correlation and values below this threshold suggesting weak correlation. The R^2 reveals the explanatory power of imputed values for true value fluctuations, where values above 0.9 signify excellent model fitting, values between 0.7 and 0.9 indicate adequate explanatory capability, and values below 0.7 reflect suboptimal fitting performance. The calculation formulas for specific parameters are listed in Table 2 below.

Table 2. Calculation formula for accuracy index.

Name of Parameter	Formula
Mean Squared Error (MSE)	$\mathrm{MSE} = \frac{1}{n}\sum_{i=1}^{n}(y_i - \hat{y}_i)^2$
Mean Absolute Error (MAE)	$\mathrm{MAE} = \frac{1}{n}\sum_{i=1}^{n}\lvert y_i - \hat{y}_i\rvert$
Pearson correlation coefficient (r)	$r = \frac{\sum_{i=1}^{n}(x_i-\overline{x})(y_i-\overline{y})}{\sqrt{\sum_{i=1}^{n}(x_i-\overline{x})^2}\cdot\sqrt{\sum_{i=1}^{n}(y_i-\overline{y})^2}}$
Coefficient of determination (R^2)	$R^2 = 1 - \frac{\sum_{i=1}^{n}(y_i-\hat{y}_i)^2}{\sum_{i=1}^{n}(y_i-\overline{y})^2}$

In the formula, n represents the number of data points, y_i is the true value, $\hat{y}_i$ is the predicted value, and $\overline{y}$ is the average value of the true values.

5.4. Short-Term Time-Scale Imputation Performance

Three symbolic variables were defined to represent submodule voltage (u_c), submodule current (i_{SM}), and submodule temperature (T_{SM}). The algorithm parameters were configured with minimum/maximum iteration limits of 50 and 1000, respectively, a termination tolerance of 1×10^{-9}, and an intentional 50% missing data ratio per dataset to simulate extreme operational conditions.

5.4.1. Bayesian Multiple Imputation for Submodule Current (i_c)

A comparison of conventional linear imputation and Bayesian multiple imputation results is depicted below Figure 9.

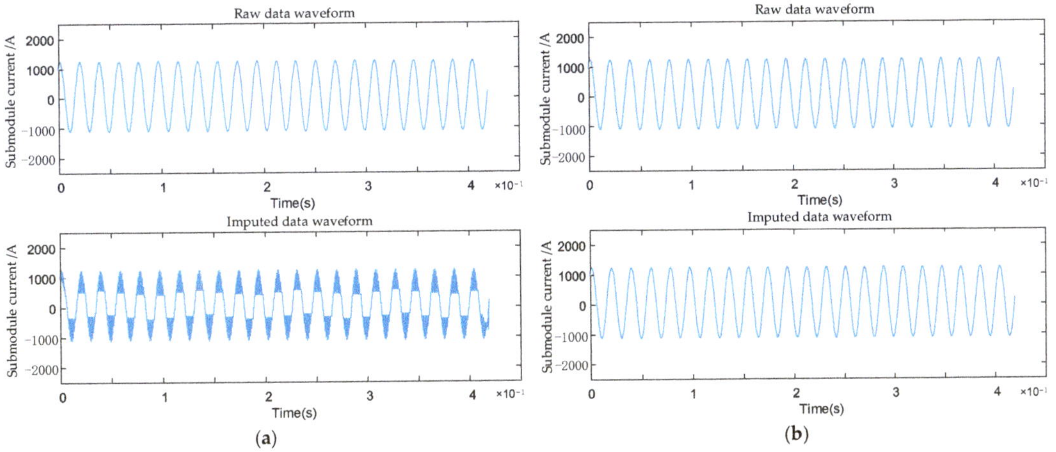

Figure 9. The interpolation results for submodule current i_c are shown in (**a**) using linear interpolation and in (**b**) employing Bayesian multiple imputation.

5.4.2. Bayesian Multiple Imputation for Submodule Voltage (u_c)

A comparative analysis of reconstruction accuracy across distinct imputation regimes is shown below Figure 10.

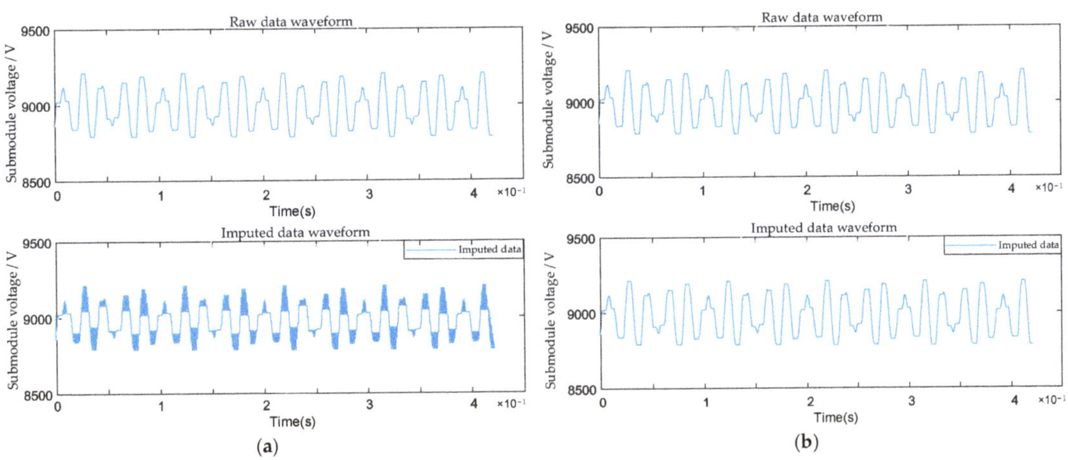

Figure 10. The interpolation results for submodule voltage u_c are shown in (**a**) using linear interpolation and in (**b**) employing Bayesian multiple imputation.

The interpolation accuracy of sub-module data is shown in Table 3.

Table 3. Sub-module data imputation accuracy.

Indicators	Submodule Current i_c		Submodule Voltage u_c	
	Linear Imputation	Multivariate Imputation	Linear Imputation	Multivariate Imputation
MSE	18,560.2328	59.6838	258.7016	9.7597
MAE	62.9265	4.9329	5.9061	1.3709
r	0.9894	1.0000	0.9937	0.9998
R^2	0.9790	1.0000	0.9874	0.9995

Under 50% missing data conditions, the simulation results in Figures 9 and 10 and associated metric calculations presented in Table 2 demonstrate that both linear and multivariate interpolation methods for submodule current (i_c) and voltage (u_c) achieve correlation coefficients (r) and R^2 values within the 0.9–1.0 range, indicating strong fitting relationships with raw data and sufficient model explanatory power, with multivariate interpolation outperforming linear interpolation. However, during MSE and MAE evaluation, linear interpolation exhibits significantly larger errors, manifesting pronounced sawtooth waveform distortions in the simulation model, demonstrating substantially lower accuracy compared to multivariate interpolation results.

5.4.3. Bayesian Multivariate Interpolation for AC-Side Single-Phase Current (i_{pa})

A comparison of conventional linear imputation and Bayesian multiple imputation results is shown below Figure 11.

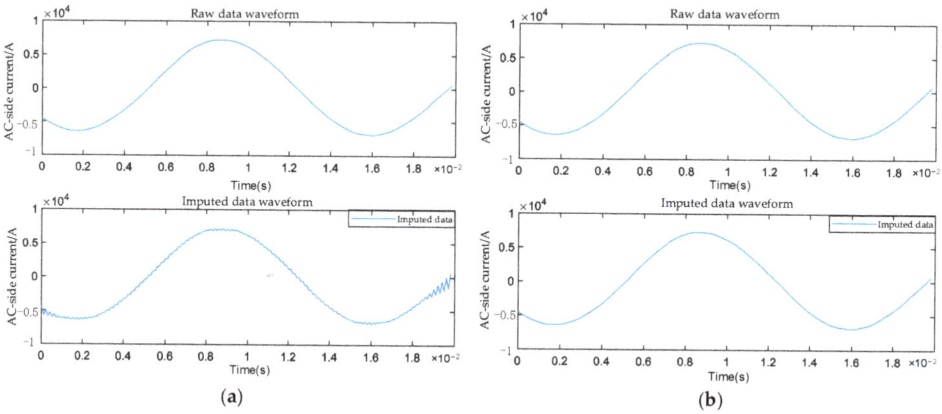

Figure 11. The interpolation results for AC-side single-phase current i_{pa} are shown in (**a**) using linear interpolation and in (**b**) employing Bayesian multiple imputation.

5.4.4. Bayesian Multivariate Interpolation for AC-Side Single-Phase Voltage (u_{sc})

A comparative analysis of reconstruction accuracy across distinct imputation regimes is shown below Figure 12.

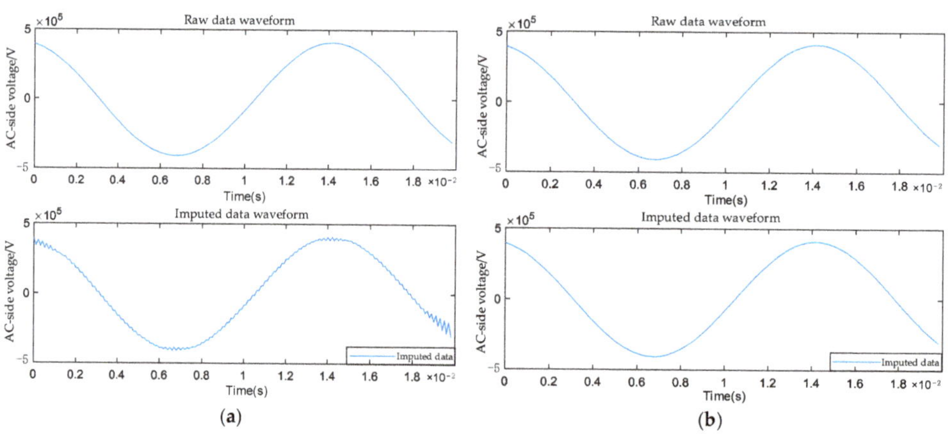

Figure 12. The interpolation results for AC-side single-phase voltage u_{sc} are shown in (**a**) using linear interpolation and in (**b**) employing Bayesian multiple imputation.

The interpolation accuracy of the bridge arm data is presented in Table 4.

Table 4. Arm data imputation accuracy.

Indicators	AC-Side-Current i_{pa}		AC-Side Voltage u_{sc}	
	Linear Imputation	Multivariate Imputation	Linear Imputation	Multivariate Imputation
MSE	91,054.7895	15.0950	2,760,691.3427	35,351.3849
MAE	151.8561	2.3564	8909.3717	118.2776
r	0.9981	1.0000	0.9983	1.0000
R^2	0.9962	1.0000	0.9967	1.0000

Under 50% missing data conditions, the simulation results from Figures 11 and 12 and corresponding metric calculations in Table 3 reveal that both linear and multivariate interpolation methods for submodule voltage and current demonstrate strong fitting performance, with correlation coefficients (r) and R^2 values within the 0.9–1.0 range, consistent with previous interpolation outcomes. The multivariate interpolation model exhibits exceptional proximity to the original data pattern due to the substantial sampling dataset. In MSE and MAE evaluations, multivariate interpolation significantly outperforms linear interpolation, achieving superior accuracy with minimal error margins.

5.5. Long-Term Timescale Imputation Performance

5.5.1. Bayesian Multivariate Interpolation for Submodule Voltage (u_c) and Current (i_c)

Under constant load conditions, the submodule voltage u_c and current i_c exhibit quasi-constant waveforms over long-term time scales due to their high switching frequency characteristics, as shown in Figure 13. However, actual UPFC field operations involve daily load fluctuations (as exemplified in telemetry data, Figure 7), granting operational telemetry data significant statistical relevance for validating analytical methodologies. This positions the daily operational dataset as a robust validation source for extended-duration performance verification in practical power electronics systems.

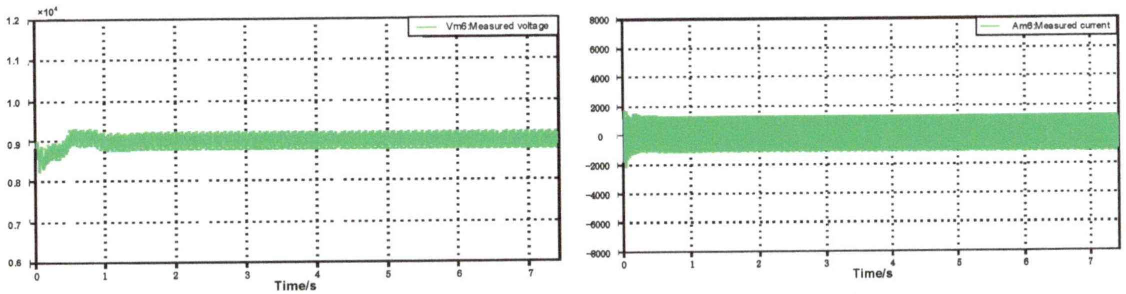

(a) (b)

Figure 13. (**a**) The operational waveform of submodule voltage u_c, and (**b**) the operational waveform of submodule current i_c.

5.5.2. Bayesian Multivariate Interpolation for Telemetry Current (i_{SM})

A comparison of the results of ordinary linear interpolation, built-in Makima interpolation, cubic spline interpolation, and Bayesian multiple imputation is shown below Figure 14.

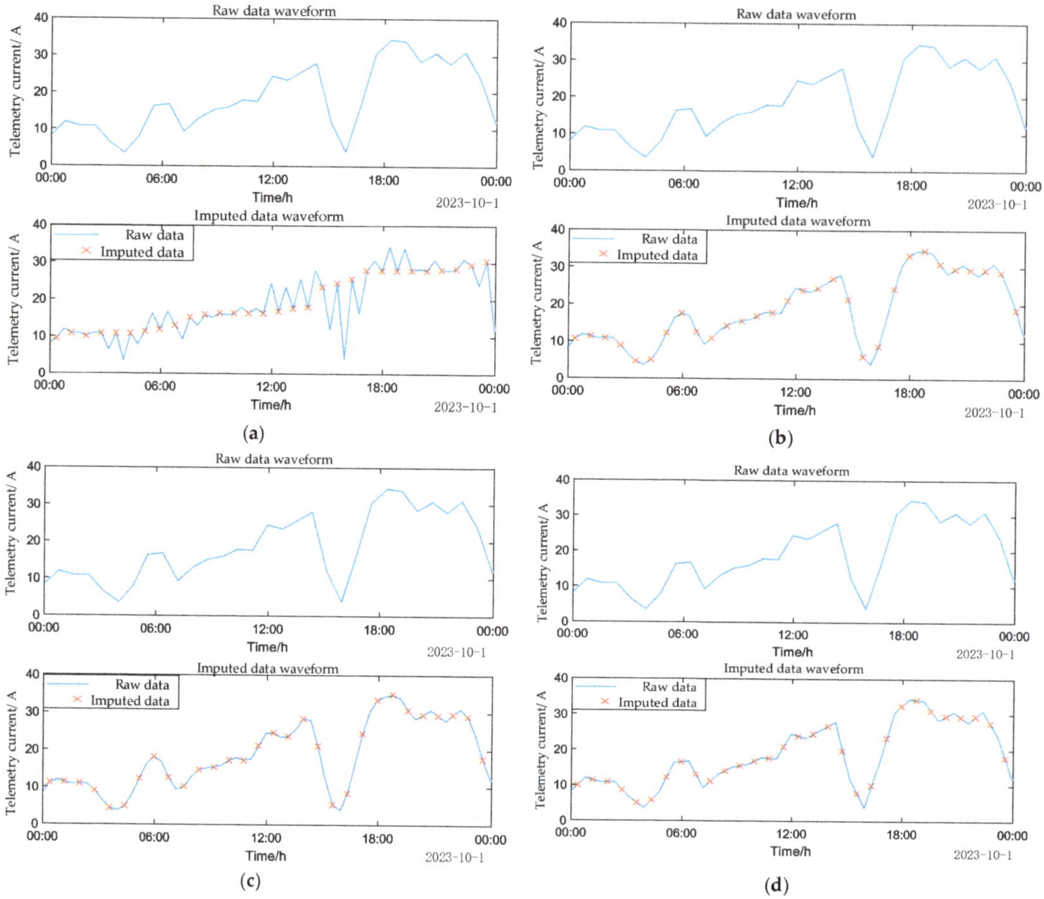

Figure 14. The interpolation results for telemetry current i_{SM} are presented in (**a**) using linear interpolation, in (**b**) with the built-in Makima interpolation, in (**c**) with the cubic spline interpolation, and (**d**) through Bayesian multiple imputation.

The interpolation accuracy of the actual telemetry current data is shown in Table 5.

Table 5. Telemetry current i_{SM} data imputation accuracy.

Indicators	Linear Imputation	Makima Imputation	Cubic Spline Interpolation	Multivariate Imputation
MSE	17.9862	0.2189	0.4407	0.1926
MAE	2.1205	0.2236	0.3588	0.2034
r	0.8791	0.9867	0.9975	1.0000
R^2	0.7727	0.9774	0.9950	1.0000

5.5.3. Bayesian Multivariate Interpolation for Telemetry Voltage (u_{SM})

A comparison of the results of ordinary linear interpolation, built-in Makima interpolation, cubic spline interpolation, and Bayesian multiple imputation is shown below Figure 15.

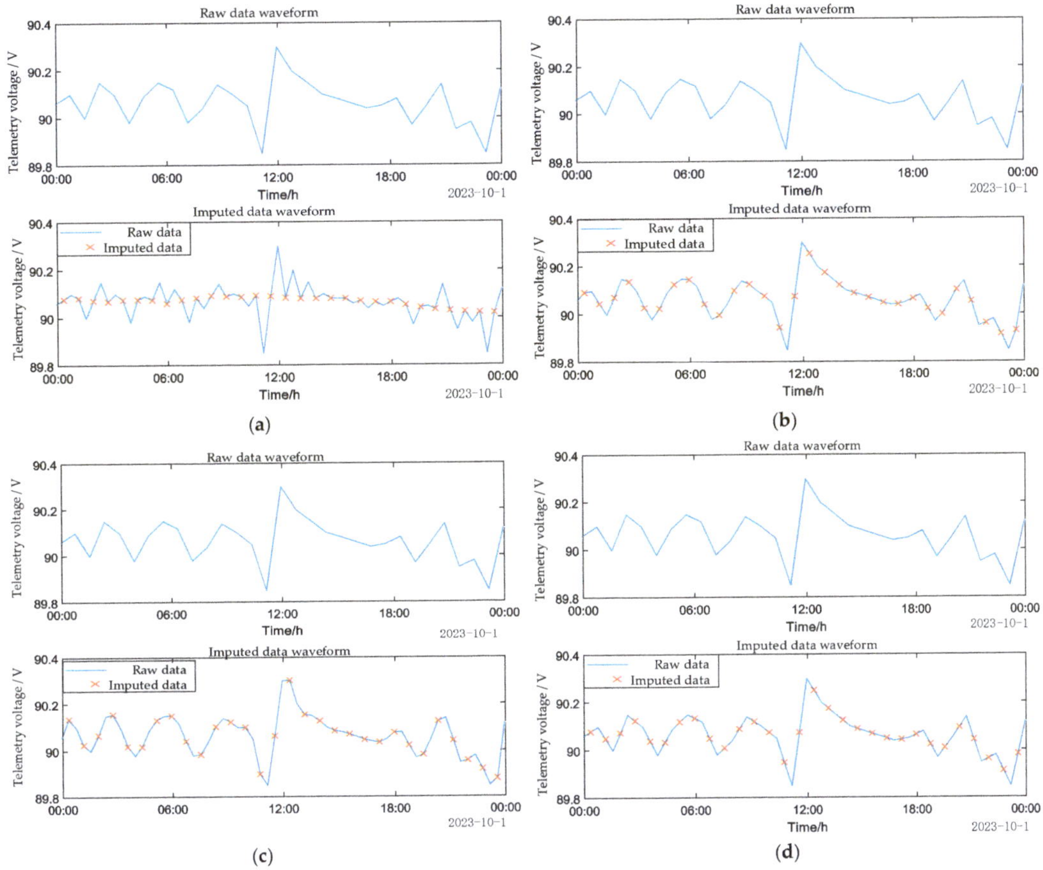

Figure 15. The interpolation results for telemetry voltage u_{SM} are presented in (**a**) using linear interpolation, in (**b**) with the built-in Makima interpolation, in (**c**) with the cubic spline interpolation, and (**d**) through Bayesian multiple imputation.

The interpolation accuracy of the actual telemetry voltage data is shown in Table 6.

Table 6. Telemetry current u_{SM} data Imputation accuracy.

Indicators	Linear Imputation	Makima Imputation	Cubic Spline Interpolation	Multivariate Imputation
MSE	0.002	0.0013	0.0004	0.0001
MAE	0.0238	0.0035	0.0099	0.0004
r	0.8329	0.9952	0.9748	1.0000
R^2	0.6936	0.9904	0.9503	1.0000

5.5.4. Bayesian Multivariate Interpolation for Telemetry Temperature (T_{SM})

A comparison of the results of ordinary linear interpolation, built-in Makima interpolation, cubic spline interpolation, and Bayesian multiple imputation is shown below Figure 16.

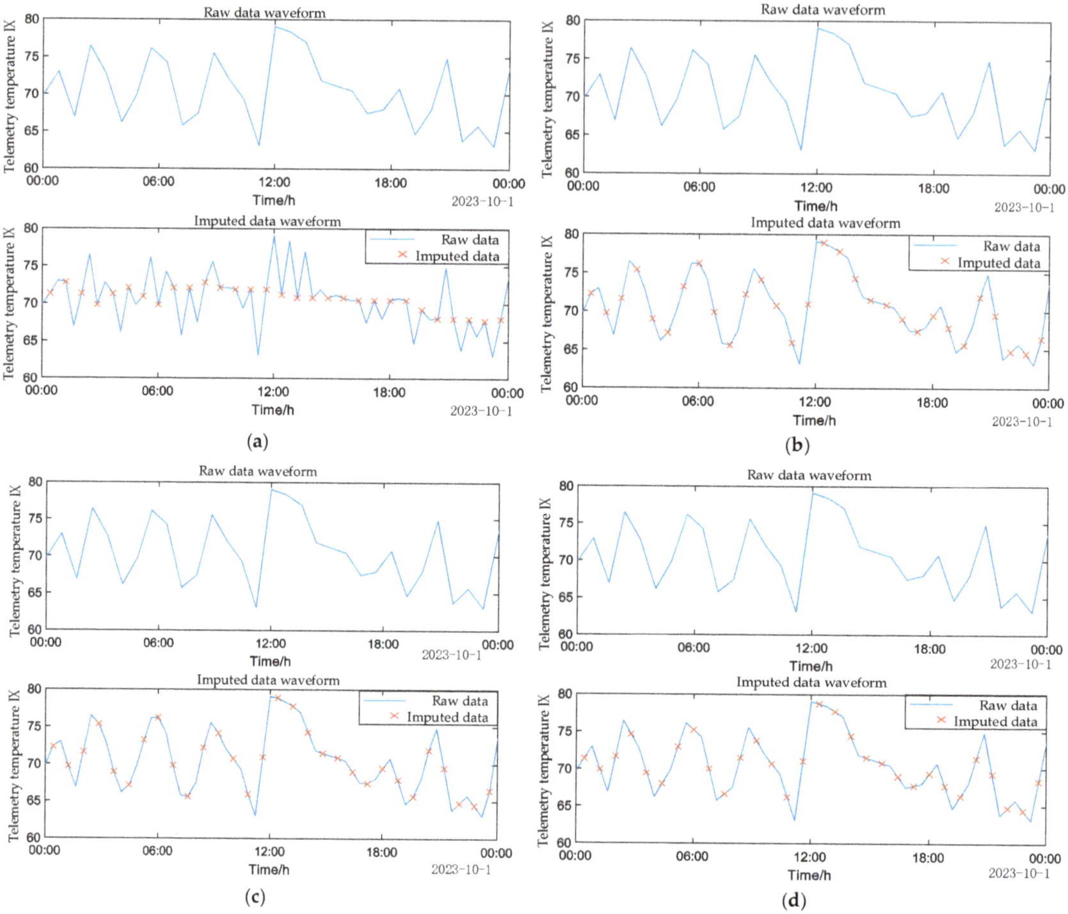

Figure 16. The interpolation results for telemetry temperature T_{SM} are presented in (**a**) using linear interpolation, in (**b**) with the built-in Makima interpolation, in (**c**) with the cubic spline interpolation, and (**d**) through Bayesian multiple imputation.

The interpolation accuracy of the telemetry temperature data is shown in Table 7.

Table 7. Telemetry temperature T_{SM} imputation accuracy.

Indicators	Linear Imputation	Makima Imputation	Cubic Spline Interpolation	Multivariate Imputation
MSE	5.5518	0.1563	0.9241	0.0141
MAE	1.3320	0.1883	0.4986	0.0167
r	0.8329	0.9958	0.9783	0.9981
R^2	0.6556	0.9916	0.9571	0.9978

Under extended-duration time frames, comparative interpolation analyses were conducted on telemetry submodule current i_{SM}, voltage u_c, and temperature T_{SM} parameters using linear imputation, Makima imputation, cubic spline interpolation, and Bayesian multivariate interpolation methods. Simulation comparisons with precision metric calculations revealed distinct performance characteristics: linear interpolation demonstrated limited effectiveness, only updating data points adjacent to missing values while introducing significant deviations from original data patterns. Makima imputation and Cubic spline interpolation achieved approximate restoration of baseline data models but exhibited partial reconstruction inaccuracies. In contrast, Bayesian multivariate interpolation precisely reconstructed all three data parameters with fidelity to original operational signatures, experimentally validating its reliability for power electronics telemetry restoration tasks.

5.6. Support Vector Machine (SVM) Hyperparameter Optimization Outcomes

The missing data ratio for three-category symbolic data was controlled at 9.4%, with confidence intervals of 63.5%, 24.4%, and 12.1%, respectively, as illustrated in Figure 17.

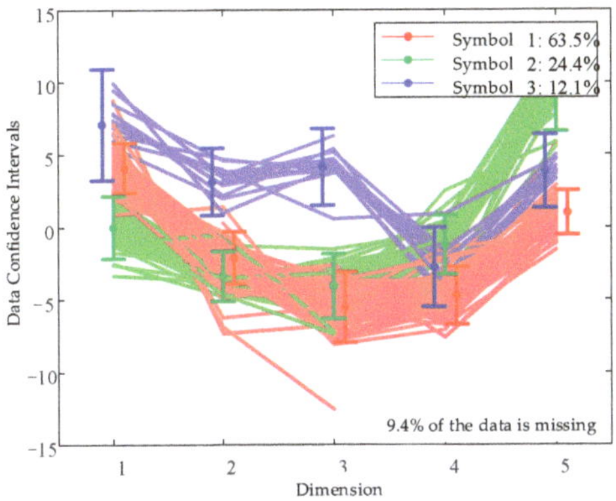

Figure 17. Model under 50% missing data condition.

In the SVM optimization section, the model results under 50% missing data conditions are shown in the figure. The proposed Bayesian multiple imputation combined with SVM methodology demonstrates controlled missing ratios (9.4% for three-category symbolic data) with confidence intervals of 63.5%, 24.4%, and 12.1%, respectively.

6. Conclusions

This study implements Bayesian multiple imputation combined with SVM optimization for UPFC submodule state information processing. Through standardized preprocessing of operational and historical submodule data, we first localize missing data points, then perform multi-stage imputation specifically tailored for short-term and long-term temporal scales of UPFC submodules, thereby validating the feasibility of Bayesian multiple imputation methodology.

Under short-term time scale validation, this study comparatively analyzed the Bayesian multiple imputation method integrated with UPFC submodule key characteristic data against conventional linear interpolation, with precision evaluated through four metrics: MSE, MAE, r, and R^2. Both methods demonstrated competent performance in data fitting and correlation preservation due to high-density sampling points, yet the conventional linear interpolation exhibited substantially higher model recovery errors owing to its reliance solely on adjacent data points. Extended missing-value imputation simulations on UPFC arm current–voltage parameters further confirmed the systematic superiority of the Bayesian method in both error suppression and fitting accuracy.

For long-term time scale verification, a comparative study was conducted across four methods: conventional linear interpolation, MAKIMA interpolation, cubic spline interpolation, and Bayesian multiple imputation. Through missing-value simulations on telemetric arm current, voltage, and temperature data, the results revealed that conventional linear interpolation failed in data fitting due to insufficient global data correlation modeling; Makima and cubic spline methods partially reconstructed data trends but retained notable residual errors; whereas the Bayesian multiple imputation achieved optimal precision in missing-value reconstruction through probabilistic dataset generation and posterior distribution modeling, outperforming other methods in both fitting fidelity and error minimization.

The methodology demonstrates universal applicability that meets the practical requirements for UPFC submodule condition monitoring. The restored dataset is subsequently integrated with SVM-based anomaly detection to identify abnormal imputation points, enabling iterative data updates that form operational datasets with enhanced estimation accuracy. The algorithm achieves improved repair precision with low computational complexity, significantly enhancing data restoration reliability while ensuring data integrity.

Author Contributions: Conceptualization, X.Y., S.S. and J.S.; methodology, Z.C.; software, S.S.; formal analysis, J.W.; investigation, C.W.; resources, C.W.; data curation, X.Y.; writing—original draft preparation, X.Y.; writing—review and editing, X.Y.; visualization, S.S.; supervision, L.Z. and C.W.; project administration, M.T.; funding acquisition, K.Z. All authors have read and agreed to the published version of the manuscript.

Funding: This project was supported by the Science and Technology Project of State Grid Jiangsu Electric Power Company: Research on Submodule-Level Digital Twin-Based Online Monitoring and Fault Location Technology for the 500 kV UPFC in Sunan (J2024023).

Data Availability Statement: The original contributions presented in this study are included in the article. Further inquiries can be directed to the corresponding author.

Conflicts of Interest: Authors Xiaoming Yu, Jun Wang, Ke Zhang, Zhijun Chen and Ming Tong were employed by the company State Grid Suzhou Power Supply Company. The remaining authors declare that the research was conducted in the absence of any commercial or financial relationships that could be construed as a potential conflict of interest.

Abbreviations

The following abbreviations are used in this manuscript:

UPFC	Unified Power Flow Controller
MMC	Modular Multilevel Converter
FACTS	Flexible AC Transmission Systems
LSTM	Long Short-Term Memory
GRU-GAN	Gated Recurrent Unit-Generative Adversarial Network
RBF	Radial basis function
SVM	Support Vector Machine
SMs	submodules
CPS-SPWM	carrier phase-shifted sinusoidal pulse width modulation
KVL	Kirchhoff's voltage law
OLS	ordinary least squares
MSE	Mean Squared Error
MAE	Mean Absolute Error
r	Pearson correlation coefficient
R^2	Coefficient of determination

References

1. Tang, Z.; Yang, Y.; Blaabjerg, F. Power electronics: The enabling technology for renewable energy integration. *CSEE J. Power Energy Syst.* **2022**, *8*, 39–52.
2. Gu, Y.; Huang, Y.; Wu, Q.; Li, C.; Zhao, H.; Zhan, Y. Isolation and Protection of the Motor-Generator Pair System for Fault Ride-Through of Renewable Energy Generation Systems. *IEEE Access* **2020**, *8*, 13251–13258. [CrossRef]

3. Elsaharty, M.A.; Luna, A.; Candela, J.I.; Rodriguez, P. A Unified Power Flow Controller Using a Power Electronics Integrated Transformer. *IEEE Trans. Power Deliv.* **2019**, *34*, 828–839. [CrossRef]
4. Lei, Y.; Li, T.; Tang, Q.; Wang, Y.; Yuan, C.; Yang, X.; Liu, Y. Comparison of UPFC, SVC and STATCOM in Improving Commutation Failure Immunity of LCC-HVDC Systems. *IEEE Access* **2020**, *8*, 135298–135307. [CrossRef]
5. Cheng, T.; Zhong, J.; Yang, B.; Wei, Y.; Xu, J.; Zhang, Z. Improved Circulation Suppression Strategy for MMC Considering Low Frequency Oscillation. In Proceedings of the 2021 11th International Conference on Power and Energy Systems (ICPES), Shanghai, China, 18–20 December 2021; pp. 57–61.
6. Huang, H.; Zhang, L.; Oghorada, O.; Mao, M. Analysis and Control of a Modular Multilevel Cascaded Converter-Based Unified Power Flow Controller. *IEEE Trans. Ind. Appl.* **2021**, *57*, 3202–3213. [CrossRef]
7. Biswas, S.; Nayak, P.K. A Fault Detection and Classification Scheme for Unified Power Flow Controller Compensated Transmission Lines Connecting Wind Farms. *IEEE Syst. J.* **2021**, *15*, 297–306. [CrossRef]
8. Ji, S.; Huang, X.; Palmer, J.; Wang, F.; Tolbert, L.M. Modular Multilevel Converter (MMC) Modeling Considering Submodule Voltage Sensor Noise. *IEEE Trans. Power Electron.* **2021**, *36*, 1215–1219. [CrossRef]
9. Dai, K.; Luo, H.; Yang, H.; He, K.; Xu, Z.; Jin, P.; Cai, H.; Xu, C. PCC Voltage Switching Harmonic Mitigation by Periodic Carrier Frequency Pulsewidth Modulation for SiC High-Frequency Active Power Filter. *IEEE J. Emerg. Sel. Top. Power Electron.* **2024**, *12*, 3040–3051. [CrossRef]
10. Wu, Q.; Zhu, Z.; Zhou, M.; Li, C.; Zhu, X. Research on Data Acquisition Method of Power System Based on Compressed Sensing Technology. In Proceedings of the 2024 2nd International Conference on Pattern Recognition, Machine Vision and Intelligent Algorithms (PRMVIA), Changsha, China, 24–27 May 2024; pp. 141–144.
11. Liu, X.; Zhang, Z. A Two-Stage Deep Autoencoder-Based Missing Data Imputation Method for Wind Farm SCADA Data. *IEEE Sens. J.* **2021**, *21*, 10933–10945. [CrossRef]
12. Liu, Y.; Duan, Z.; Hu, P.; Hu, C.; Zheng, J.; Li, Z. MMC Submodule Topology With CVSB and DCFRT Capability for Offshore Wind Farms Black-Start. *IEEE Trans. Ind. Appl.* **2024**, *60*, 2891–2900. [CrossRef]
13. Zhou, K.; Gao, F.; Zhang, Y.; Hou, Z.; Nie, K.; Xu, T.; Wang, H.; Xu, K. Missing Data Imputation for Photovoltaic Station Based on Spatiotemporal Features. In Proceedings of the 2024 CPSS & IEEE International Symposium on Energy Storage and Conversion (ISESC), Xi'an, China, 8–11 November 2024; pp. 355–359.
14. Yang, Y.; Wang, W.; Zhong, Y.; Jin, C.; Zhao, C.; Qiu, B.; Jia, W. Short-Term EVs Charging Load Prediction Under Low-quality Data Environment Based on GRU-GAN-GOA-MKELM Framework. In Proceedings of the 2024 International Seminar on Artificial Intelligence, Computer Technology and Control Engineering (ACTCE), Wuhan, China, 28–29 September 2024; pp. 145–152.
15. Kim, J.; Lee, J. Statistical Classification of Vehicle Interior Sound Through Upsampling-Based Augmentation and Correction Using 1D CNN and LSTM. *IEEE Access* **2022**, *10*, 100615–100626. [CrossRef]
16. Lee, S.-Y.; Connerton, T.P.; Lee, Y.-W.; Kim, D.; Kim, D.; Kim, J.-H. Semi-GAN: An Improved GAN-Based Missing Data Imputation Method for the Semiconductor Industry. *IEEE Access* **2022**, *10*, 72328–72338. [CrossRef]
17. Allella, F.; Chiodo, E.; Giannuzzi, G.M.; Lauria, D.; Mottola, F. On-Line Estimation Assessment of Power Systems Inertia With High Penetration of Renewable Generation. *IEEE Access* **2020**, *8*, 62689–62697. [CrossRef]
18. Zhao, Z.; Luo, X.; Wu, J.; Xie, J.; Gong, S.; Ni, Q.; Lai, C.S.; Lai, L.L. Model Reduction for Grid-Forming Hybrid Renewable Energy Microgrid Clusters Based on Multi-Timescale Characterization. *IEEE Trans. Smart Grid* **2024**, *15*, 1227–1242. [CrossRef]
19. Liu, H.; Hu, F.; Su, J.; Wei, X.; Qin, R. Comparisons on Kalman-Filter-Based Dynamic State Estimation Algorithms of Power Systems. *IEEE Access* **2020**, *8*, 51035–51043. [CrossRef]
20. Ahmed, H.M.; Abdulrazak, B.; Blanchet, F.G.; Aloulou, H.; Mokhtari, M. Long Gaps Missing IoT Sensors Time Series Data Imputation: A Bayesian Gaussian Approach. *IEEE Access* **2022**, *10*, 116107–116119. [CrossRef]
21. Lei, M.; Labbe, A.; Wu, Y.; Sun, L. Bayesian Kernelized Matrix Factorization for Spatiotemporal Traffic Data Imputation and Kriging. *IEEE Trans. Intell. Transp. Syst.* **2022**, *23*, 18962–18974. [CrossRef]
22. Sun, P.; Liu, X.; Lin, M.; Wang, J.; Jiang, T.; Wang, Y. Transmission Line Fault Diagnosis Method Based on Improved Multiple SVM Model. *IEEE Access* **2023**, *11*, 133825–133834. [CrossRef]
23. Gong, X.; Zou, B.; Duan, Y.; Xu, J.; Luo, Q.; Yang, Y. Multiple Kernel SVM Based on Two-Stage Learning. *IEEE Access* **2020**, *8*, 101133–101144. [CrossRef]

energies

MDPI

Article

A Novel Wide-Gain-Range Variable-Structure DC/DC Converter Based on an LLC Resonant Converter

Qingqing He, Shun Tang, Dan Ren, Zhaoyang Tang, Qisheng Zhu, Chao Tang and Keliang Zhou *

The School of Automation, Wuhan University of Technology, Wuhan 430070, China;
heqingqing@whut.edu.cn (Q.H.); shuntang7@whut.edu.cn (S.T.); 363470@whut.edu.cn (D.R.);
trigo@whut.edu.cn (Z.T.); sleepzhu@whut.edu.cn (Q.Z.); tang_chao@whut.edu.cn (C.T.)
* Correspondence: 12956@whut.edu.cn

Abstract

The LLC resonant converter, as an isolated DC-DC conversion topology, has been widely adopted in industrial applications. However, when operating under wide input/output voltage ranges, a broad switching frequency range is required to achieve the desired voltage gain. This wide frequency variation complicates the design of magnetic components, causes loss of soft-switching characteristics, and deteriorates electromagnetic interference (EMI) performance. To address these challenges, this paper presents a detailed analysis of the L-LCLC resonant converter. By controlling the connection/disconnection of additional inductors and capacitors through switching devices, the topology achieves structural reconfiguration to enhance the voltage gain range. Optimal mode transition points are selected to ensure stable operation during mode transitions, thereby reducing design complexity, minimizing transition losses, and suppressing voltage/current stress. The parameter design methodology for the additional reactive components is systematically developed. The converter's performance is validated with Simulink, and the experimental prototype is established with 100 W. Both simulation and experimental results confirm that the L-LCLC resonant converter achieves a wide voltage gain range within a narrow frequency band while maintaining stable mode transitions.

Keywords: variable-structure resonant converter; DC/DC; L-LCLC resonant converter

1. Introduction

The LLC resonant converter has gained widespread adoption as one of the most prevalent isolated DC-DC converter topologies, owing to its high efficiency, high power density, and excellent soft-switching characteristics [1–4]. This topology finds extensive applications in critical domains, including on-board battery chargers (OBCs), server and telecommunication power supplies, industrial automation equipment, medical power supplies, LED lighting drivers, and renewable energy systems [5–7]. Figure 1 illustrates the circuit configuration of a conventional full-bridge LLC resonant converter incorporating a full-bridge rectifier stage. In LLC resonant converters, the power transfer mechanism is inherently facilitated through the resonant circuit (comprising inductor L_r and capacitor C_r) and the isolation transformer. As schematically depicted in Figure 1, the application of a square-wave voltage waveform at the V_{AB} node establishes a sinusoidal current profile within the resonant L_r–C_r network. This alternating current is subsequently propagated to the secondary side through magnetic coupling in the transformer, enabling energy delivery to the load [8–10]. The voltage regulation mechanism in an LLC resonant converter is realized by modulating the switching frequency of switches Q_1–Q_4. This modulation dynamically alters the impedance division ratio within the resonant tank, thereby controlling the output voltage [11]. This operational principle enables precise control of the voltage conversion gain, thereby maintaining a stabilized output voltage U_o under varying load conditions.

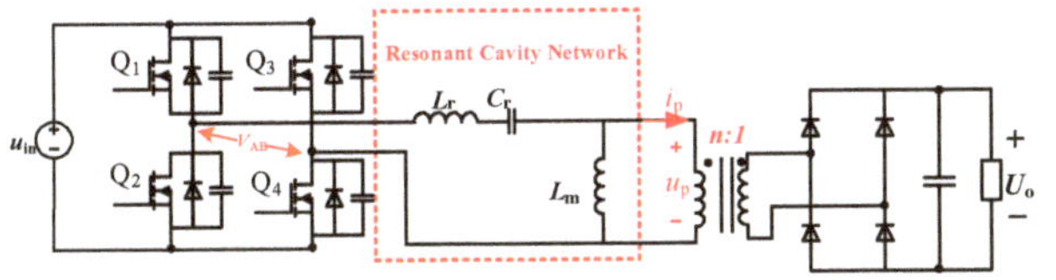

Figure 1. Full-bridge LLC resonant converter circuit configuration.

Pulse frequency modulation (PFM) is widely used in LLC resonant converters, adjusting the switching frequency to modulate resonant tank impedance [12]. However, above resonance, PFM-controlled LLC converters exhibit reduced voltage gain sensitivity at higher frequencies, complicating wide voltage-range applications (e.g., on-board chargers and solar systems) [13,14]. These scenarios require excessively broad frequency ranges to meet strict gain specifications, increasing magnetics complexity, compromising zero-voltage switching, and degrading EMI performance [15–18]. To address these limitations while achieving extended voltage gain within narrow frequency bands, numerous research approaches have been proposed from different technical perspectives.

Based on the modification locations, LLC resonant converters can be categorized into three distinct subcategories: primary-side inverters, resonant tanks, and secondary-side rectifiers. In [19], a novel topology was introduced featuring an additional switch integrated into the primary full-bridge inverter, dual independent resonant tanks (L_r/C_r), transformers, and full-wave rectifiers. This configuration enables multi-mode operation through primary switch state regulation, significantly expanding the achievable voltage gain range. However, the incorporation of redundant components, particularly magnetic elements, substantially increases circuit complexity and manufacturing costs while compromising power density. In [20,21], different basic topologies, such as Buck–Boost and Buck, are cascaded with an LLC resonant converter to achieve extremely high voltage conversion ratios. However, this approach suffers from significant drawbacks, including highly complex system control and potentially slow dynamic response.

A fractional magnetizing inductance with a parallel-connected auxiliary switch was presented in [22]. By actuating the auxiliary switch, the resonant tank configuration can be dynamically reconfigured to operate in different states. Various secondary-side rectification schemes have been proposed in [23–27]. Notably, ref. [27] replaced the conventional full-wave rectifier with a reconfigurable voltage-doubler rectifier capable of operating in Type-4, Type-5, and Type-6 configurations. Nevertheless, the achievable gain ranges of existing rectification topologies remain constrained, which inevitably expands the required frequency variation range and consequently limits potential application scenarios.

To enable the proposed converter to achieve a wide voltage gain range within a narrow frequency band, this paper introduces a multi-element resonant converter methodology by incorporating one or two additional inductors/capacitors into the conventional LLC resonant tank structure. This approach generates multiple resonant frequencies in the system transfer function, thereby enhancing the gain characteristics. Through comparative analysis of the gain profiles across these derived LLC resonant converter topologies, the configuration with optimal gain performance is selected for detailed investigation. The resonant tank configurations of the proposed LLC-derived topologies are illustrated in Figure 2.

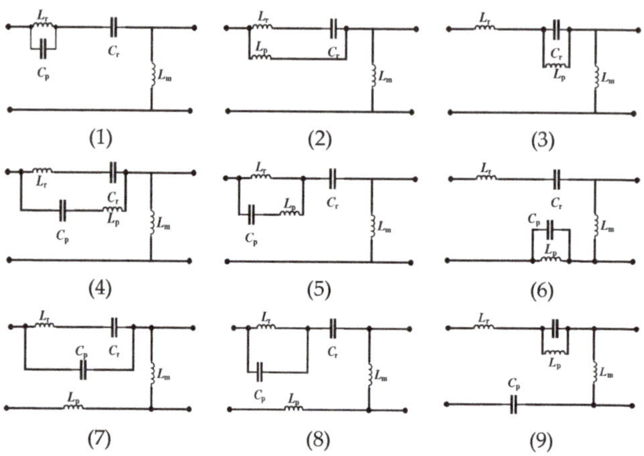

Figure 2. Resonant tank structure of the derivative resonant converter.

Through systematic modeling and parametric analysis of the voltage gain characteristics for LLC derivative topologies 1 through 9, the gain expression for each topology is established based on the Fundamental Harmonic Approximation (FHA) method. Subsequently, their dynamic voltage gain curves as a function of normalized frequency are plotted using a consistent set of parameters. Building upon this, a comparative analysis is conducted by plotting the gain curves of the derivative topologies alongside those of the conventional LLC resonant converter on the same set of axes. The results are illustrated in Figure 3, where the red curves represent the gain of the derivative topologies and the blue curve corresponds to that of the conventional LLC converter.

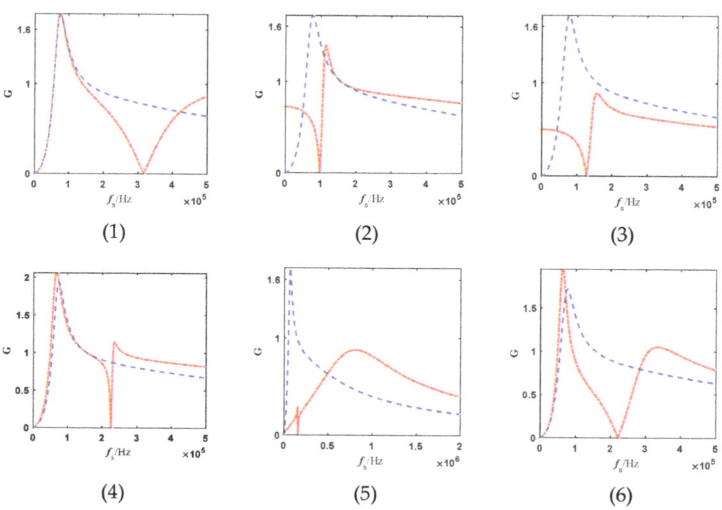

Figure 3. *Cont.*

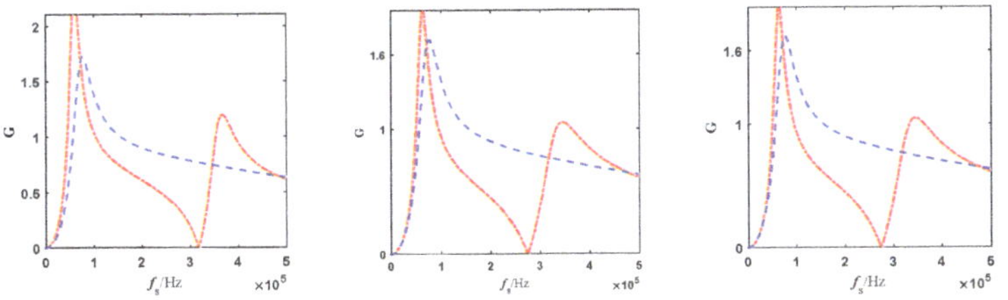

(7) (8) (9)

Figure 3. Voltage gain curves of the derivative resonant converters.

Image analysis reveals that Topology 4 demonstrates significantly superior gain range compared to other configurations. This ultra-steep gain characteristic enables stable high-frequency output under wide input voltage scenarios while allowing substantial voltage regulation through minimal frequency adjustments. Such performance effectively reduces switching device stress, making it particularly suitable for high-dynamic applications. Consequently, Topology 4 is selected as the primary research subject and designated as the L-LCLC resonant converter in this study.

By controlling the connection/disconnection of additional inductors and capacitors through switching devices, the L-LCLC resonant converter achieves structural reconfiguration to enhance the voltage gain range. Compared to the LLC topology, the L-LCLC configuration requires additional reactive components but achieves superior gain characteristics. To optimize overall system efficiency, a hybrid control strategy is implemented: the converter operates in L-LCLC mode when the switching frequency exceeds the resonant frequency, and transitions to LLC mode under sub-resonant conditions. The mode transition occurs precisely at unity voltage gain, where both topologies share identical operating frequencies and gain values. This design methodology effectively minimizes voltage/current stress during mode transitions while maintaining high efficiency across the entire operational range.

The contributions of this study are summarized as follows: (1) This paper presents a detailed analysis of the voltage gain characteristics for the L-LCLC resonant converter, accompanied by a complete derivation of the governing mathematical equations. Furthermore, the soft-switching operational regions of the converter are systematically delineated through a thorough steady-state analysis. Based on these findings, a systematic parameter design methodology is also proposed. (2) A hybrid mode transition strategy is implemented to improve overall system efficiency. The converter seamlessly switches between LLC and L-LCLC modes at the unity gain point, where both topologies share identical operating frequencies and gain characteristics. This design effectively mitigates transient voltage and current surges during mode transitions while maintaining high efficiency across the entire operational range.

The organization of this paper is structured as follows: Section 1 reviews key challenges in LLC resonant converters and explains the motivation for the L-LCLC topology. Section 2 provides a detailed analysis of the L-LCLC converter's operational principles, covering gain characteristics, soft-switching properties, and hybrid mode transitions. Section 3 proposes a design methodology for critical components, highlighting optimization of the parallel inductor L_p and capacitor C_p. Section 4 validates theoretical analysis via simulations, while Section 5 confirms performance through experimental testing on a 100 W prototype. Section 6 summarizes contributions and practical implications.

2. Principal Analysis of Converter

2.1. The Equivalent Model of the Circuit and the Gain

Figure 4 illustrates the circuit configuration of the L-LCLC resonant converter. Compared to the conventional LLC topology, this configuration incorporates an additional inductor L_p and capacitor C_p within the resonant cavity network.

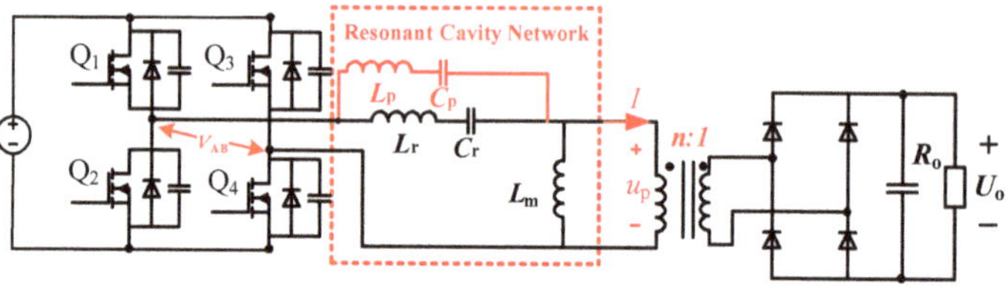

Figure 4. L-LCLC resonant converter circuit configuration.

The L-LCLC resonant converter is modeled using the fundamental harmonic approximation method. By linearizing the operational characteristics of the high-frequency transformer and composite resonant tank network at the fundamental frequency, an AC equivalent circuit model incorporating core resonant components is established. The simplified equivalent circuit diagram of the L-LCLC resonant converter is illustrated in Figure 5. In Figure 5, E_i represents the input voltage to the resonant tank, while I_p, I_r, and I_m are the currents flowing through capacitor C_p, capacitor C_r, and inductor L_m, respectively. E_o denotes the overall output voltage of the resonant converter.

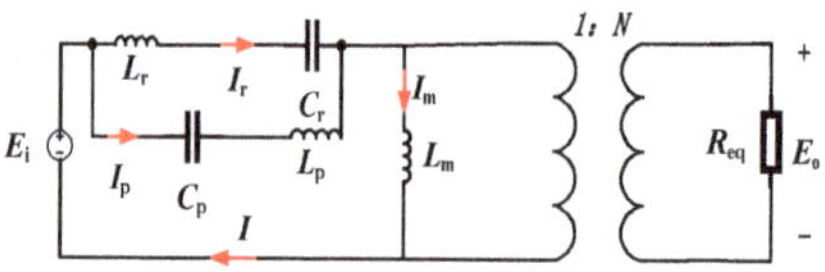

Figure 5. L-LCLC resonant converter equivalent circuit.

Based on the fundamental harmonic analysis method, the secondary-side equivalent resistance R_{eq} of the transformer can be derived as

$$R_{eq} = \frac{8R_o}{\pi^2} \tag{1}$$

By formulating the KVL/KCL, the system characteristics can be derived as

$$\begin{cases} E_i(s) = I_p(s)(\frac{1}{sC_p} + sL_p) + \frac{E_o(s)}{N} \\ I(s) = I_p(s) + I_r(s) \\ I(s) = \frac{E_o(s)}{NsI_m} + \frac{NE_o(s)}{R_{eq}} \\ I_p(s)(\frac{1}{sC_p} + sL_p) = I_r(s)(\frac{1}{sC_r} + sL_r) \end{cases} \tag{2}$$

By solving the aforementioned system of equations, the following expressions can be derived as

$$\frac{E_i(s)}{E_o(s)} = \frac{1}{N} + (\frac{N}{R_{eq}} + \frac{1}{NsL_m})\frac{(1+s^2L_pC_p)(1+s^2L_rC_r)}{s(C_r+C_p)+s^3(L_r+L_p)C_rC_p} \tag{3}$$

The reciprocal of the L-LCLC resonant converter's voltage gain in the frequency domain can be formulated as

$$\begin{aligned}\frac{1}{G} = \ & 1+\frac{1}{\omega L_m}\frac{(1-\omega^2 L_p C_p)(1-\omega^2 L_r C_r)}{\omega^3(L_r+L_p)C_r C_p-\omega(C_r+C_p)} \\ & +j\frac{(1-\omega^2 L_p C_p)(1-\omega^2 L_r C_r)}{R_{ac}(\omega^3(L_r+L_p)C_r C_p-\omega(C_r+C_p))}\end{aligned} \tag{4}$$

The L-LCLC resonant converter exhibits three distinct resonant frequencies $f_1 = 1/2\pi\sqrt{L_r C_r}, f_2 = 1/2\pi\sqrt{L_p C_p}, f_3 = \sqrt{(L_p+C_p)}/2\pi\sqrt{C_r C_p(L_p+C_p)}$. The converter's gain profile is defined by three critical resonant frequencies. At switching frequencies f_1 and f_2, the converter achieves a unity voltage gain. This occurs because series resonance—between L_r and C_r at f_1, and between L_p and C_p at f_2—causes the equivalent impedance of the resonant branch to approach zero, allowing the output voltage to approximate the input voltage. When operating at f_3, the entire L_r-C_r-L_p-C_p network transitions into parallel resonance, creating an infinite equivalent impedance across the resonant branch. This state effectively blocks energy transfer to the output, and the converter gain is approximately zero.

2.2. Soft-Switching Characteristics Analysis

As derived from Figure 5, the input impedance Z_{in} of the L-LCLC resonant network can be expressed as

$$Z_{in} = \frac{\omega^2 {L_m}^2 R_{ac}}{{R_{ac}}^2+\omega^2 {L_m}^2}+j(\frac{{R_{ac}}^2 \omega L_m}{{R_{ac}}^2+\omega^2 {L_m}^2}+\frac{\omega^4 L_r L_p C_r C_p - L_r C_r \omega^2 - L_p \omega^2 C_p + 1}{\omega^3(L_r+L_p)C_r C_p - \omega C_p - \omega C_r}) \tag{5}$$

The normalized frequencies are defined as f_{n1}, f_{n2}, and f_{n3}, where $R_{ac} = R_{eq}n^2$, $f_{n1} = f_s/f_1, f_{n2} = f_s/f_2, f_{n3} = f_s/f_3$, and where f_s denotes the switching frequency. The structural voltage gain G can be expressed as

$$G(f_{n1}) = \frac{1}{\sqrt{(1+\frac{P_2 P_3}{P_1 \lambda_1 f_{n1}})^2+(T\frac{P_2 P_3}{P_1})^2}} \tag{6}$$

where $T = \sqrt{(Lr/Cr)}/R_{ac}$, $P_1 = (\lambda_1\lambda_3 + \lambda_1\lambda_3)/(\lambda_2 + \lambda_2\lambda_3)f_{n1}^2 - 1$, $P_2 = 1 - \lambda_1\lambda_3 f_{n1}^2/\lambda_2$, $P_3 = (1 - f_{n1}^2)/f_{n1}(1 + \lambda_3)$, $\lambda_1 = L_m/L_r$, $\lambda_2 = L_m/L_p$, $\lambda_3 = C_p/C_r$.

When the input impedance Z_{in} exhibits purely resistive characteristics, the voltage gain becomes inherently resistive and can be derived as

$$G = \sqrt{\frac{(\lambda_1 {f_{n1}}^2 + \lambda_1\lambda_3 {f_{n1}}^2)P_1}{(\lambda_1 {f_{n1}}^2 + \lambda_1\lambda_3 {f_{n1}}^2)P_1 + (1-{f_{n1}}^2)P_2}} \tag{7}$$

Based on Equations (6) and (7), the inherently resistive voltage gain curves and parametric gain curves for varying T values can be plotted, as illustrated in Figure 6. In the Figure 6, the number 1 represents the blue area, number 2 represents the yellow area, and number 3 represents the pink area.

The operational zones of the L-LCLC resonant converter are categorized into three distinct regions, defined by the purely resistive characteristic curve and the boundary line $f_{n1} = 1$:

Region 1: Located to the left of both the boundary line $f_{n1} = 1$ and the purely resistive characteristic curve, the converter exhibits capacitive behavior and operates in the ZCS mode.

Region 2: Positioned to the left of the boundary line $f_{n1} = 1$ and to the right of the purely resistive characteristic curve, the converter exhibits inductive behavior. Here, the voltage gain $G > 1$, indicating a step-up operation mode, with the system functioning in the ZVS regime.

Region 3: Positioned to the right of the boundary line $f_{n1} = 1$ and to the left of the purely resistive characteristic curve, where the voltage gain $G < 1$ indicates a step-down operation mode, with the system operating in the ZVS regime.

The operational analysis of the L-LCLC resonant converter focuses on switching frequencies below f_3, as frequencies exceeding f_3 are excluded from practical design considerations. To prioritize ZVS operation, the converter should be designed to operate in Region 3, as established in the preceding analysis.

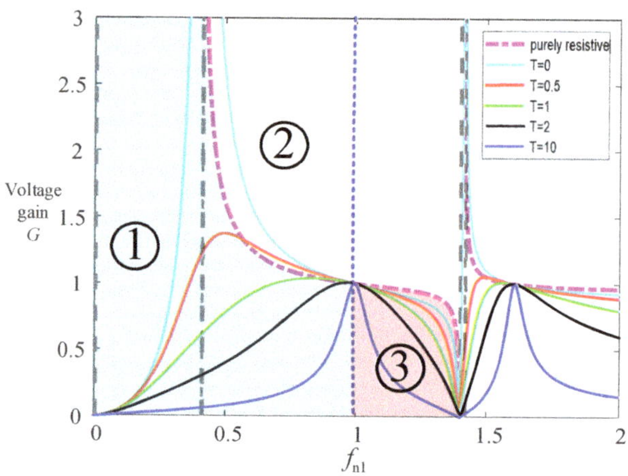

Figure 6. The voltage gain curve of the L-LCLC resonant converter.

2.3. *Mode Transition Strategy*

As illustrated in Figure 7, the voltage gain curves of the two resonant converters reveal distinct operational characteristics. Under the LLC frequency-modulation (FM) mode, the converter operates in a step-up configuration, whereas the L-LCLC structure-variable frequency modulation mode enables step-down operation. Since a mode switch is required, we should select an appropriate switching point.

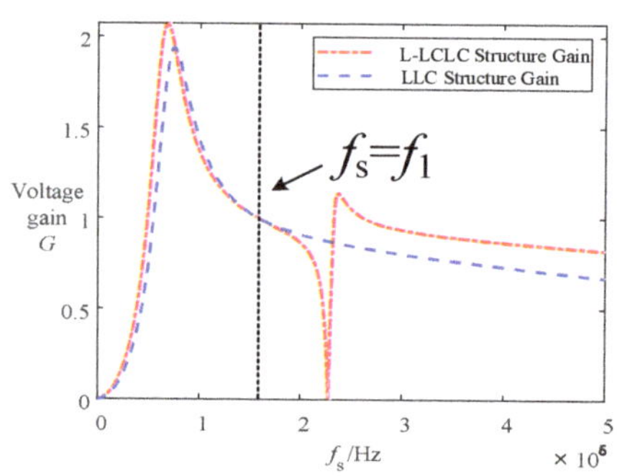

Figure 7. Composite gain curve graph.

Two conventional methods exist for determining mode transition points. The first is based on equal efficiency between operational modes, requiring loss calculations (conduction/switching) in both modes, but faces computational complexity and numerical instability. The second prioritizes resonant frequency proximity, balancing voltage gain stability and efficiency sensitivity to frequency deviations. In this study, both converter modes operate under frequency modulation control and share a common resonant frequency point. At this specific switching frequency, both modes achieve unity voltage gain with zero frequency deviation, significantly mitigating voltage/current stress during mode transitions. Therefore, this resonant frequency point is selected as the optimal mode transition point.

3. Parameter Design

Extensive research has been conducted by scholars on the circuit parameter design of LLC resonant converters [28–32]. Building upon these foundations, this paper presents a systematic parameter design methodology for the additional parallel-connected inductor and capacitor introduced in the L-LCLC resonant converter topology.

Based on the hybrid control strategy definition, the LLC FM mode operates under minimum input voltage conditions, achieving the maximum voltage gain $G_{\max}$. Conversely, the L-LCLC FM mode functions at maximum input voltage, yielding the minimum voltage gain $G_{\min}$. With constant output voltage maintained, the following relationships hold:

$$V_{in\min}G_{\max} = V_{in\max}G_{\min} = nV_o \tag{8}$$

As established in the operational characteristics, the LLC frequency modulation mode operates in the step-up region ($G > 1$), whereas the L-LCLC structure-variable FM mode functions in the step-down region ($G < 1$). Under these conditions, the voltage gain relationship can be derived as

$$G_{\max} \geq 1 \tag{9}$$

$$G_{\min} \leq 1 \tag{10}$$

The peak gain, denoted as G_{peak}, is fully utilized by setting $G_{\text{peak0}} = G_{\max}$. To ensure design robustness, a 20% safety margin is incorporated. The available peak gain G_{peak0} can thus be calculated as

$$G_{peak0} = G_{\max} = \frac{G_{peak}}{120\%} \tag{11}$$

To ensure the converter operates in the LLC FM mode when the switching frequency is below f_{n1}, and achieves steeper voltage gain curves beyond the mode transition point compared to the LLC FM mode, the composite parallel resonant frequency of the L_r-C_r-L_p-C_p network must exceed the series resonant frequency of the L_r-C_r branch. This condition guarantees optimal performance across both operational modes.

$$f_3 > f_1 \tag{12}$$

To prevent the voltage gain from collapsing to zero, the composite parallel resonant frequency of the L_r-C_r-L_p-C_p network should be designed to exceed the converter's maximum switching frequency $f_{s\max}$. This condition ensures operational stability across the entire frequency modulation range while maintaining inherent voltage regulation capability.

$$f_{s\max} < f_3 \tag{13}$$

The parametric design methodology initiates with selecting an appropriate λ_3 value within the typical operational range of 0.1–0.5 for the L-LCLC resonant converter operating in regulated output mode. Based on Equation (6), different λ_2 values are sequentially evaluated to satisfy the predefined gain constraints, thereby establishing the feasible λ_2 range. Within this range, the optimal λ_2 parameter is determined according to specific gain sensitivity requirements. Following λ_2 selection, an inverse iterative process is implemented: maintaining the chosen λ_2 while re-evaluating λ_3 variations under identical gain constraints to determine its optimized range. The final λ_2/λ_3 parameters are then consolidated through this bidirectional optimization process. Substituting these optimized values into Equations (14) and (15) subsequently yields the concrete specifications for L_p and C_p components. The systematic parameter selection flowchart is illustrated in Figure 8.

$$L_p = \frac{L_m}{\lambda_2} \tag{14}$$

$$C_p = \lambda_3 C_r \tag{15}$$

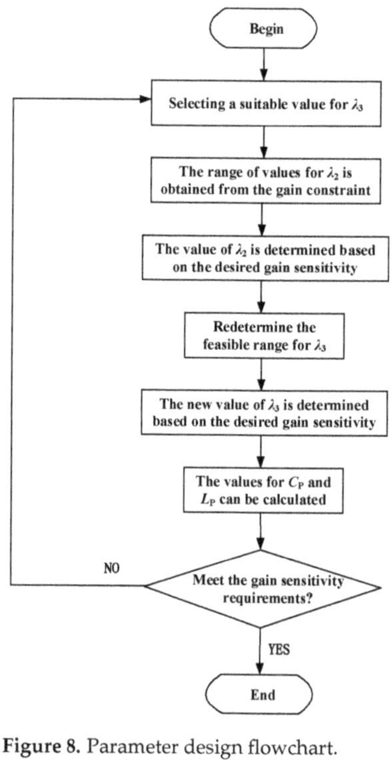

Figure 8. Parameter design flowchart.

Taking the design process presented in this paper as an example, the design procedure is initiated by setting λ_3 = 0.2, followed by evaluating multiple λ_2 values to generate the gain profiles illustrated in Figure 9a. The optimal λ_2 value of 2.6 is determined based on the specified gain sensitivity requirements. Subsequently, an iterative refinement process adjusts λ_3 variations while maintaining the selected λ_2, producing the gain characteristics shown in Figure 9b. Through comparative analysis of these gain profiles, λ_3 is finalized at 0.25, considering both sensitivity optimization and operational margin. The inductance L_p and capacitance C_p parameters are then calculated by substituting the optimized λ_2/λ_3 set into Equations (14) and (15). The final design parameters of the converter, including the optimized L_p and C_p values, are systematically summarized in Table 1.

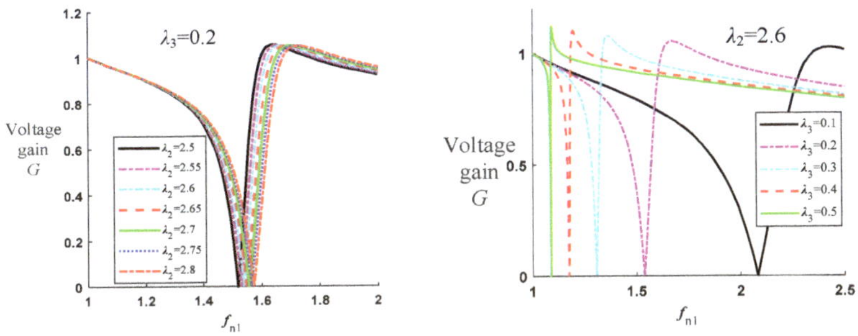

(**a**) Voltage Gain Curve of the Converter (1) (**b**) Voltage Gain Curve of the Converter (2)

Figure 9. Voltage gain curves of the L-LCLC resonant converter with parametric variations in λ2 and λ3.

Table 1. Prototype specifications.

Specifications	Value
Input voltage V_{in}	140~250 V
Rated output voltage V_{out}	200 V
Rated power P	100 W
Resonant inductance L_r	108.85 μH
Resonant capacitor C_r	9.09 nF
Resonant inductance L_p	167.47 μH
Resonant capacitor C_p	2.27 nF
Excitation inductance L_m	435.41 μH
Transformer ratio n (1: N)	1

4. Simulation Study

To validate the feasibility of the proposed methodology, a Simulink (MATLAB R2024a version)-based simulation study is first conducted. The simulation was performed using a discrete solver with a sample time of 5 ns. The parasitic capacitance of each primary-side switching device was set to 1.5 nF. The switching frequency is regulated by a Proportional–Integral (PI) controller to maintain a stable output voltage. Furthermore, the engagement of the L_p and C_p branches is dynamically controlled. This is achieved by continuously sampling the input and output voltages to calculate the real-time converter gain. Based on this gain, the branch is disconnected from the circuit when the gain is greater than one and reconnected when the gain is less than one. The key parameters of the converter, including component values and operational specifications, are presented in Table 1.

Figure 10a,b present the simulated waveforms of the capacitor C_r current I_{cr} and output voltage illustrating two operational scenarios: Figure 10a input voltage U_{in} = 160 ± 20 V with regulated output voltage U_o = 200 V. From top to bottom, the waveforms represent input voltage, output voltage, and capacitor current. As shown in Figure 10b, when U_{in} decreases from 180 V to 140 V, the output voltage maintains stable regulation around 100 V. Similarly, when U_{in} increases from 210 V to 250 V, the output voltage remains tightly regulated at 100 V. This demonstrates excellent voltage regulation stability across wide input voltage variations, enabling the converter to adapt to applications requiring different voltage levels.

Figure 11a,b present the voltage, current, and power waveforms of primary-side switching devices, along with gate trigger voltage signals, at switching frequencies of 150 kHz and 220 kHz, respectively. As observed in both subfigures, during switch turn-on transitions, no overlap occurs between voltage and current waveforms, resulting in zero turn-on loss. This confirms the successful realization of ZVS, which aligns precisely with theoretical predictions.

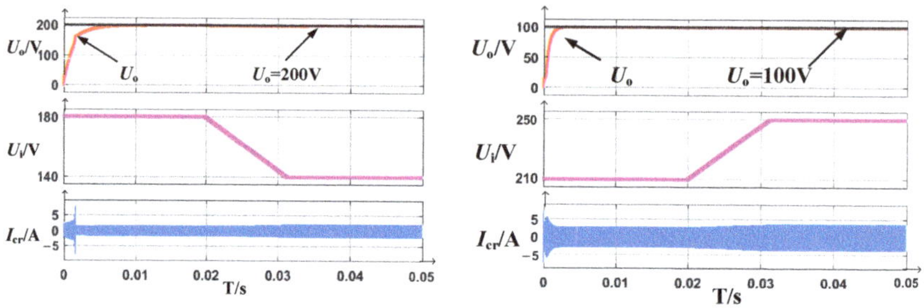

(**a**) (**b**)

Figure 10. Simulated waveforms of the capacitor C_r current I_{cr} and output voltage. (**a**) Dynamic curve graph showing the reduction of input voltage from 180 V to 140 V; (**b**) Dynamic curve graph showing the increase in input voltage from 210 V to 250 V.

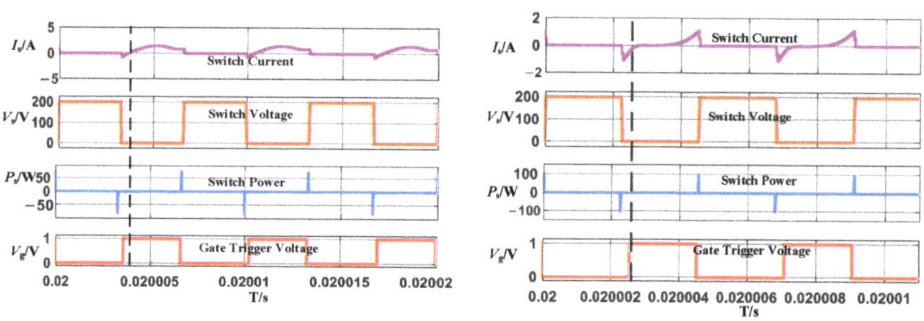

(**a**) 150 KHz (**b**) 220 KHz

Figure 11. Simulation curves of soft-switching characteristics for L-LCLC resonant converter.

5. Experimental Result

A 100 W experimental prototype is established to verify parameter design. The specifications of the experimental prototype are consistent with Table 1. Figure 12 shows the photograph of the experimental prototype.

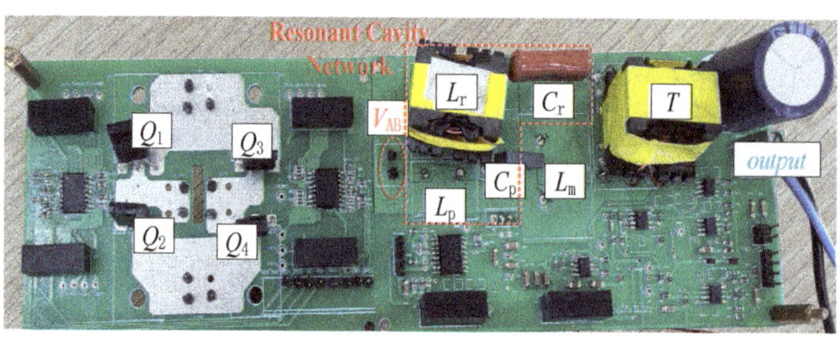

Figure 12. The prototype of the converter.

Figure 13 presents a comparative analysis of voltage gain characteristics between the LLC and L-LCLC resonant converter modes under a 200 V input voltage. Figure 13a,b illustrate the experimental waveforms of square-wave input voltage V_{AB} and output voltage at a switching frequency of 160 kHz for LLC and L-LCLC modes, respectively. Figure 13c,d display corresponding results at 228 kHz. At 160 kHz, both modes exhibit output voltage fluctuations around 200 V. However, when operating at 228 kHz, the LLC mode produces an output voltage of approximately 130 V, while the L-LCLC mode achieves a significantly lower output voltage of 40 V. This demonstrates the L-LCLC configuration's ability to extend the voltage gain range by over 300% compared to the conventional LLC topology, validating its superior performance in wide-gain applications.

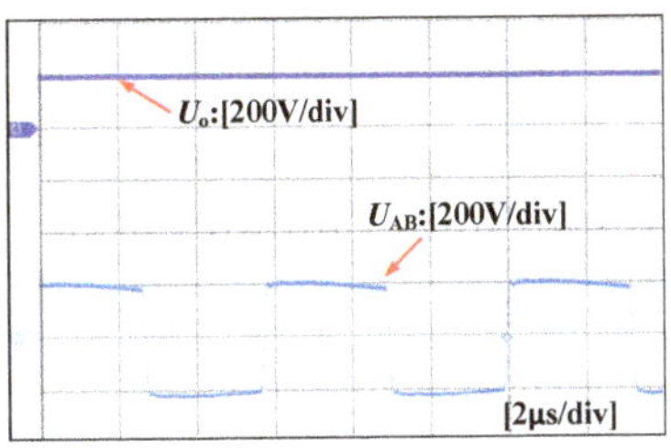

(**a**) Voltage Gain Curves (1)

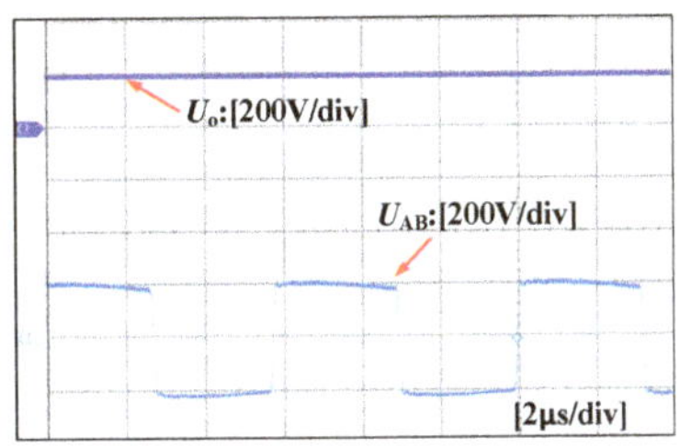

(**b**) Voltage Gain Curves (2)

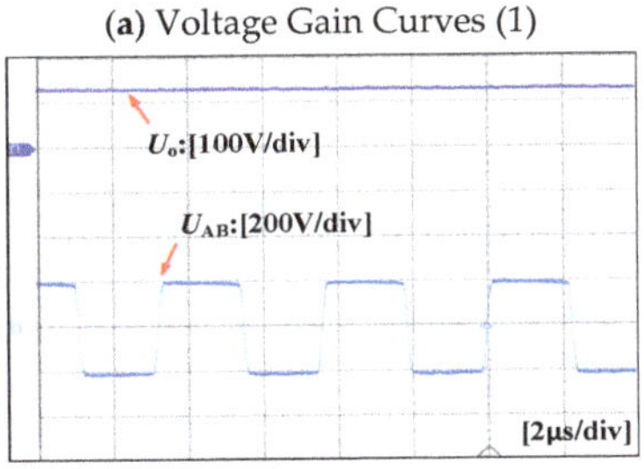

(**c**) Voltage Gain Curves (3)

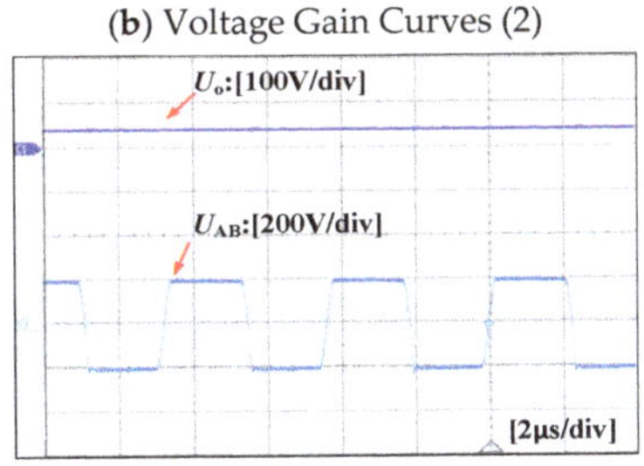

(**d**) Voltage Gain Curves (4)

Figure 13. Comparative voltage gain simulation results of the LLC and L-LCLC resonant converters.

Figure 14a,b illustrate the experimental waveforms of the current I and output voltage under an input voltage of 200 V and switching frequency of 160 kHz, corresponding to output powers of 100 W and 75 W, respectively. From top to bottom, the waveforms represent output voltage, current I, square-wave input voltage V_{AB}, and gate-source voltage of the switches. As shown in the left figure, at full load (100 W), the output voltage remains stably regulated around 200 V. Similarly, the right figure demonstrates that even at reduced load (75 W), the output voltage maintains tight regulation at approximately 200 V. This confirms the converter's robust voltage regulation capability across varying load conditions, enabling its application in scenarios with diverse power requirements.

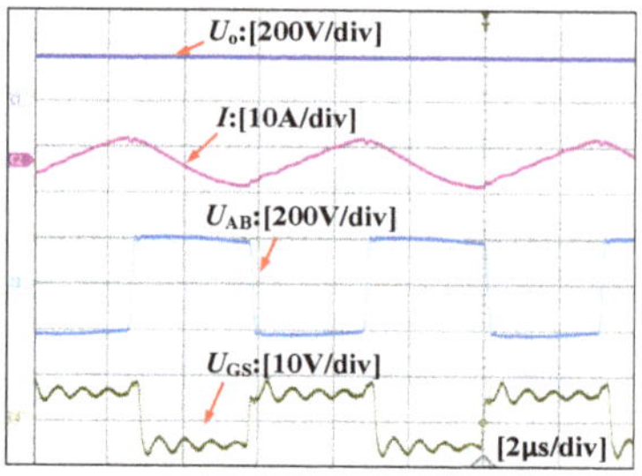

(**a**) Output power: 100 W

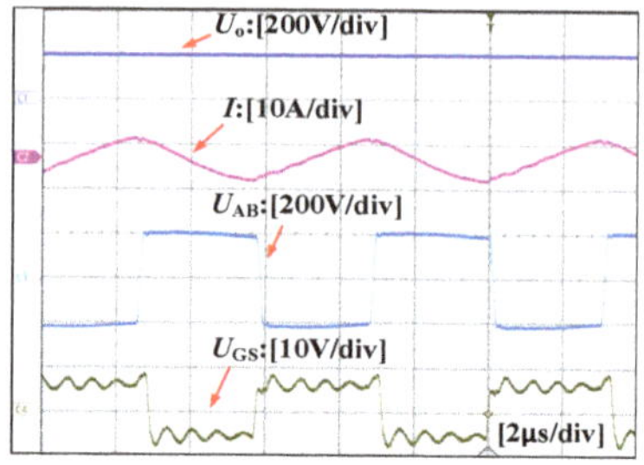

(**b**) Output power: 75 W

Figure 14. Experimental Characteristics of L-LCLC Resonant Converter under Load Variations.

As shown in Figure 15, during the transition from LLC to L-LCLC mode, the output voltage remains stably regulated around 200 V, demonstrating excellent transient stability. This confirms the effectiveness of the hybrid control strategy in maintaining output regulation during structural reconfiguration.

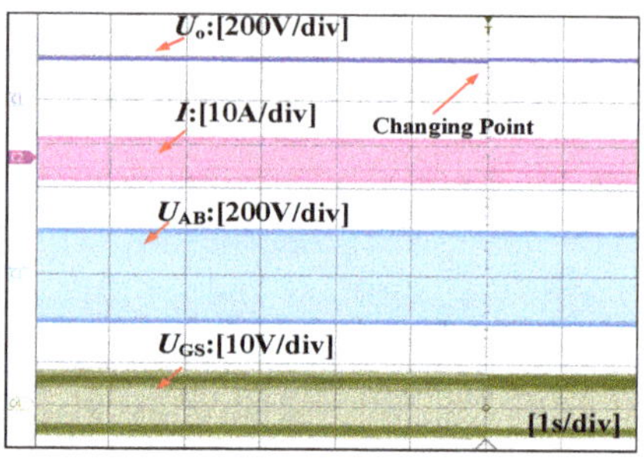

Figure 15. Transition process from LLC mode to L-LCLC mode.

6. Conclusions

This paper presents a variable-structure LLC resonant converter where an auxiliary switching network enables dynamic configuration transition. By controlling the conduction state of additional switches, the converter operates in LLC FM mode when the voltage gain exceeds unity, and switches to L-LCLC structural-variable FM mode under sub-unity gain conditions. A systematic design methodology for the parallel-connected inductors and capacitors in the L-LCLC configuration is proposed. Simulation and experimental results demonstrate robust output voltage stability at 200 V under input voltage step changes (180 V → 140 V and 210 V → 250 V) and varying load conditions (100 W and 75 W). The L-LCLC topology achieves a wide voltage gain range from 1 to 0.2 within a narrow switching frequency band of 160–228 kHz, compared to the conventional LLC converter's gain range of 0.65–1 under identical operational conditions. The converter has a wider input-voltage range with a narrow switching frequency range.

Author Contributions: Conceptualization, methodology, data curation, and writing, Q.H., S.T., D.R., C.T. and Q.Z.; analysis, software, and experiments, Q.H., S.T., Z.T. and K.Z. All authors have read and agreed to the published version of the manuscript.

Funding: This research was funded by the National Natural Science Foundation of China (52307219) and the Wuhan Science and Technology Bureau Project (2023020201020307).

Data Availability Statement: Data are contained within the article.

Conflicts of Interest: The author declares no conflicts of interest.

References

1. Gupta, P.P.; Kumar, N.; Nangia, U. Design and Analysis of LLC Resonant Converter and CCCV Topology for Battery Charging. In Proceedings of the 2022 2nd Asian Conference on Innovation in Technology (ASIANCON), Ravet, India, 26–28 August 2022; pp. 1–5.
2. Wang, Q.; Wang, W.; Shuai, Z.; Peng, Y. Condition Monitoring for Resonant Capacitors in LLC Resonant Converter. In Proceedings of the 2024 IEEE 10th International Power Electronics and Motion Control Conference (IPEMC2024-ECCE Asia), Chengdu, China, 17–20 May 2024; pp. 3545–3550.
3. Wu, H.; Mu, T.; Gao, X.; Xing, Y. A secondary-side phase-shift controlled LLC resonant converter with reduced conduction loss at normal operation for hold-up time compensation application. *IEEE Trans. Power Electron.* **2015**, *30*, 5352–5357. [CrossRef]
4. Sathya, K.; Guruswamy, K.P. Performance Analysis for LLC Resonant Converter in Electric Vehicle Applications. In Proceedings of the 2023 International Conference on Recent Trends in Electronics and Communication (ICRTEC), Mysore, India, 10–11 February 2023; pp. 1–6.
5. Nabih, A.; Li, Q. Design of 98.8% Efficient 400-to-48-V LLC Converter with Optimized Matrix Transformer and Matrix Inductor. *IEEE Trans. Power Electron.* **2023**, *38*, 7207–7225. [CrossRef]
6. Zhang, J.; Du, X.; Qian, C.; Du, R.; Hu, X.; Tai, H.-M. Thermal Management of IGBT Module in the Wind Power Converter Based on the ROI. *IEEE Trans. Ind. Electron.* **2022**, *69*, 8513–8523. [CrossRef]
7. He, Y.; Jia, Y.; Gao, L.; Xue, W.; Yang, Y.; Gao, T. A Control Method for Current Sharing in Interleaved LLC Resonant Converter. In Proceedings of the 2024 IEEE 7th International Conference on Electronic Information and Communication Technology (ICEICT), Xi'an, China, 31 July–2 August 2024; pp. 641–645.
8. Chen, Z.; Shi, W.; Wang, K. Research on New Wide-input LLC Resonant Converter. In Proceedings of the 2023 IEEE 6th International Electrical and Energy Conference (CIEEC), Hefei, China, 12–14 May 2023; pp. 2145–2149.
9. Wei, Y.; Luo, Q.; Chen, S.; Sun, P.; Altin, N. Comparison among different analysis methodologies for LLC resonant converter. *IET Power Electron.* **2019**, *12*, 2236–2244. [CrossRef]
10. Vuchev, A.S.; Grigorova, T.G.; Vuchev, S.A. Theoretical Analysis of an LLC Resonant DC-DC Converter at Boost Mode. In Proceedings of the 2023 XXXII International Scientific Conference Electronics (ET), Sozopol, Bulgaria, 13–15 September 2023; pp. 1–4.
11. Guan, Y.; Li, X.; Wang, W.; Li, B.; Yang, W.; Wang, Y.; Xu, D. A High-Performance 3:1 Conversion Ratio DC–DC Converter: Analysis Method and Modular Adoption. *IEEE Trans. Power Electron.* **2024**, *39*, 4412–4425. [CrossRef]
12. Wang, Y.; Gao, Z.; Dou, S. A Novel Resonant DC-DC Converter with Wide Voltage Gain. In Proceedings of the 2023 China Automation Congress (CAC), Chongqing, China, 17–19 November 2023; pp. 1349–1353.
13. Menke, M.F.; Avila, S.L. Multi-objective Genetic Algorithm Based Optimal Design Methodology for LLC Resonant Converter. In Proceedings of the 2022 14th Seminar on Power Electronics and Control (SEPOC), Santa Maria, Brazil, 12–15 November 2022; pp. 1–6.
14. Yang, X.; Sha, D. Automatic Current Limit Strategy for LLC DC-DC Converter for Overload Operation. *IEEE Trans. Power Electron.* **2024**, *39*, 9917–9928. [CrossRef]
15. Vuchev, A.; Grigorova, T.; Vuchev, S. Analysis of Continuous Current Mode of an LLC Resonant DC-DC Converter at ZVS. In Proceedings of the 2022 13th National Conference with International Participation (ELECTRONICA), Sofia, Bulgaria, 19–20 May 2022; pp. 1–4.
16. Zhang, J.; Shen, H.; Sun, H.; Wang, Z. Review on the Thermal Models Applications in the Reliability of Power Semiconductor Device. *IET Power Electron.* **2025**, *18*, e70065. [CrossRef]
17. Zhang, J.; Shen, H.; Du, X.; Chen, R. Condition Monitoring the Inhomogeneous Thermal Fatigue of Multichip IGBT Module Based on the Thermal Attenuation Coefficient. *IEEE Trans. Power Electron.* **2025**, *40*, 2114–2125. [CrossRef]
18. Barbi, I. Analysis of Oscillation Between Transformer Leakage Inductance and Schottky Diode Capacitance in DC–DC Converters. *IEEE Trans. Power Electron.* **2024**, *39*, 9127–9131. [CrossRef]
19. Li, C.; Zhou, M.; Wang, H. An H5-Bridge-Based Asymmetric LLC Resonant Converter with an Ultrawide Output Voltage Range. *IEEE Trans. Ind. Electron.* **2020**, *67*, 9503–9514. [CrossRef]
20. Liu, Q.; Qian, Q.; Ren, B.; Xu, S.; Sun, W.; Yang, L. A Two-Stage Buck–Boost Integrated LLC Converter with Extended ZVS Range and Reduced Conduction Loss for High-Frequency and High-Efficiency Applications. *IEEE J. Emerg. Sel. Top. Power Electron.* **2021**, *9*, 727–743. [CrossRef]
21. Wang, W.; Liu, Y.; Zhao, J.; Zhang, P.; Loh, P.C. A Dynamic Control Method for Buck + LLC Cascaded Converter with a Wide Input Voltage Range. *IEEE Trans. Power Electron.* **2023**, *38*, 1522–1534. [CrossRef]

22. Hu, H.; Fang, X.; Chen, F.; Shen, Z.J.; Batarseh, I. A Modified High-Efficiency LLC Converter with Two Transformers for Wide Input-Voltage Range Applications. *IEEE Trans. Power Electron.* **2013**, *28*, 1946–1960. [CrossRef]
23. Wen, H.; Jiao, D.; Lai, J.-S.; Strydom, J.; Lu, B. A MHz LCLCL Resonant Converter Based Single-Stage Soft-Switching Isolated Inverter with Variable Frequency Modulation. *IEEE Trans. Power Electron.* **2022**, *37*, 10797–10807. [CrossRef]
24. Alaql, F.; Alhatlani, A.; Batarseh, I. Multi-Mode Rectifier-Based LLC Resonant Converter for Wide Input Voltage Range Applications. In Proceedings of the 2021 IEEE Applied Power Electronics Conference and Exposition (APEC), Phoenix, AZ, USA, 14–17 June 2021; pp. 349–354.
25. Shang, M.; Wang, H. A Voltage Quadrupler Rectifier Based Pulsewidth Modulated LLC Converter with Wide Output Range. *IEEE Trans. Ind. Appl.* **2018**, *54*, 6159–6168. [CrossRef]
26. Alaql, F.; Rezaii, R.; Gullu, S.; Elrais, M.T.; Batarseh, I. A Switchable Rectifier-based LLC Resonant Converter for Photovoltaic Applications. In Proceedings of the 2021 IEEE Energy Conversion Congress and Exposition (ECCE), Singapore, 24–27 May 2021; pp. 2093–2098.
27. Shang, M.; Wang, H.; Cao, Q. Reconfigurable LLC Topology with Squeezed Frequency Span for High-Voltage Bus-Based Photovoltaic Systems. *IEEE Trans. Power Electron.* **2018**, *33*, 3688–3692. [CrossRef]
28. Chen, W.; Xu, M.; Ge, Y.; Guo, Q.; Jiang, T.; Che, X. Parameter Design of High Power LLC Resonant Converter. In Proceedings of the 2022 IEEE 5th International Conference on Electronics Technology (ICET), Chengdu, China, 13–16 May 2022; pp. 421–425.
29. Yuan, Y.; Xu, L.; Yao, C.; Yuan, J.; Wang, X.; Ruan, X. Simplified Time-Domain Design Methodology for LLC Resonant Converter in EV Charging Station Application. In Proceedings of the 2024 3rd International Conference on Power Systems and Electrical Technology (PSET), Tokyo, Japan, 5–8 August 2024; pp. 332–337.
30. Hou, Z.; Jiao, D.; Lai, J.-S. An Ultrawide Range Pulse Width Modulated LLC Converter with Voltage Multiplier Rectifiers. *IEEE Trans. Power Electron.* **2025**, *40*, 6216–6229. [CrossRef]
31. Zhou, X.; Chen, C.; Ni, K.; Mi, H.; Yang, J.; Yang, T. A Parameter Design Method for L-LLC Resonant Converter. In Proceedings of the 2023 IEEE 2nd International Power Electronics and Application Symposium (PEAS), Guangzhou, China, 10–13 November 2023; pp. 1205–1210.
32. Wei, Y.; Pereira, T.; Pan, Y.; Liserre, M.; Blaabjerg, F.; Mantooth, H.A. RMS Current based Automated Optimal Design Tool for LLC Resonant Converters. In Proceedings of the 2022 IEEE Applied Power Electronics Conference and Exposition (APEC), Houston, TX, USA, 20–24 March 2022; pp. 2146–2153.

energies

MDPI

Article

Analysis and Design of a Multiple-Driver Power Supply Based on a High-Frequency AC Bus

Qingqing He, Zhaoyang Tang, Wenzhe Zhao and Keliang Zhou *

School of Automation, Wuhan University of Technology, Wuhan 430070, China; heqingqing@whut.edu.cn (Q.H.); trgio@whut.edu.cn (Z.T.); 330444@whut.edu.cn (W.Z.)
* Correspondence: keliang.zhou@whut.edu.cn

Abstract

Multi-channel LED drivers are crucial for high-power lighting applications. Maintaining a constant average forward current is essential for stable LED luminous intensity, necessitating drivers capable of consistent current delivery across wide operating ranges. Meanwhile, achieving precise current sharing among channels without incurring high costs and system complexity is a significant challenge. Leveraging the constant-current characteristics of the LCL-T network, this paper presents a multi-channel DC/DC LED driver comprising a full-bridge inverter, a transformer, and a passive resonant rectifier. The driver generates a high-frequency AC bus with series-connected diode rectifiers, a structure that guarantees excellent current sharing among all output channels using only a single control loop. Fully considering the impact of higher harmonics, this paper derives an exact solution for the output current. A step-by-step parameter design methodology ensures soft switching and enhanced switch utilization. Finally, experimental verification was conducted using a prototype with five channels and 200 W, confirming the correctness and accuracy of the theoretical analysis. The experimental results showed that within a wide input voltage range of 380 V to 420 V, the driver was able to provide a stable current of 700 mA to each channel, and the system could achieve a peak efficiency of up to 94.4%.

Keywords: parameter design; DC-supplied LED drivers; full-bridge inverter; high-frequency AC bus; LCL-T network

1. Introduction

With the rapid developments in semiconductor lighting, light-emitting diodes (LEDs) have been extensively applied to lighting equipment and consumer electronic products because of their high luminous efficiency, pollution-free characteristics, and super longevity [1,2]. To achieve sufficient lumens in the aforementioned applications, LEDs are coupled in series and parallel [3]. This configuration yields the requirement of multi-output constant current for LED drivers. It also requires power factor correction (PFC) and electrical isolation to satisfy power quality standards and safety regulations [4,5].

At the present time, approaches for realizing multi-channel constant-current output can be broadly categorized into single-converter and multi-converter schemes [6]. The multi-converter approach, wherein each LED load is driven by a dedicated DC/DC converter, offers flexible control but is often compromised by high system cost, volume, and complexity [7]. Consequently, single-converter multiple-output topologies have been extensively investigated. A key technical challenge for these topologies lies in achieving inherent current balancing among the parallel output branches.

As seen in Figure 1, this two-stage LED driver represents a common option for medium- to high-power applications. PFC converters make up the first stage, and DC/DC converters make up the second stage. Boost, buck, or other converters are often adopted in the first stage to obtain a high power factor and constant output voltage [8–11]. To address this challenge, this paper focuses on constant-current control multi-output DC/DC LED drivers [12,13] and aims to develop low-cost, high-power density, and high-efficiency solutions [14].

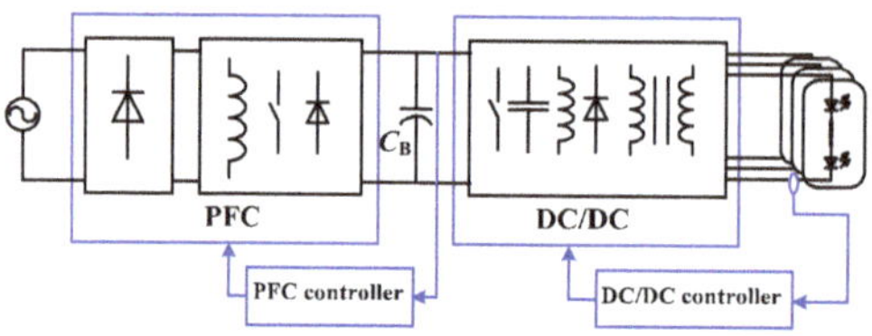

Figure 1. The typical two-stage multi-output LED driver.

Among existing single-converter multiple-output current-balancing techniques, passive methods, such as those employing resistors or capacitors, feature a simple structure but suffer from poor regulation accuracy and introduce additional power losses [15]. In contrast, active methods, while capable of enhancing accuracy by monitoring and controlling each output channel, necessitate complex control circuitry and a significant increase in component count [16]. In recent years, advanced topologies based on resonant converters, single-inductor multiple-output (SIMO) structures, and electrolytic capacitor-less designs have been successively proposed, effectively enhancing system performance [17–20]. In particular, schemes that utilize the constant-current characteristics of resonant networks, such as the LCL-T topology, to establish a high-frequency AC bus have demonstrated considerable potential for multi-load applications [21].

Despite some progress in recent years, developing solutions that achieve high-precision current sharing, require minimal components, and feature simple control schemes remains a major challenge. This paper addresses these issues by proposing a novel multi-channel LED driver based on a high-frequency AC bus. The driver employs an innovative series rectifier architecture to achieve inherent current sharing.

A current-sharing scheme of a resonant network is proposed in [21]. Due to the LCL-T constant-current properties, the rectifiers can be designed to function similarly to current sources. In order to provide multiple constant-current outputs, multiple LCL-T rectifiers are coupled to the same high-frequency AC voltage bus. In order to simplify the circuit, increase power density, and reduce costs, more components should be shared among different outputs. The port for the output of the LCL-T passive resonant network can be used to build a high-frequency AC bus in accordance with the circuit duality concept. Each output can also share the passive resonant network apart from the original full-bridge inverter and high-frequency isolation transformer. Thus, as seen in Figure 2, the constant-current multi-output LED driver constructed using a high-frequency AC bus can be built. The focus of this paper is the second-stage DC/DC converter, whose input, U_{dc}, is provided by a front-end PFC circuit. Figure 2 shows the detailed topology of the proposed DC/DC converter part.

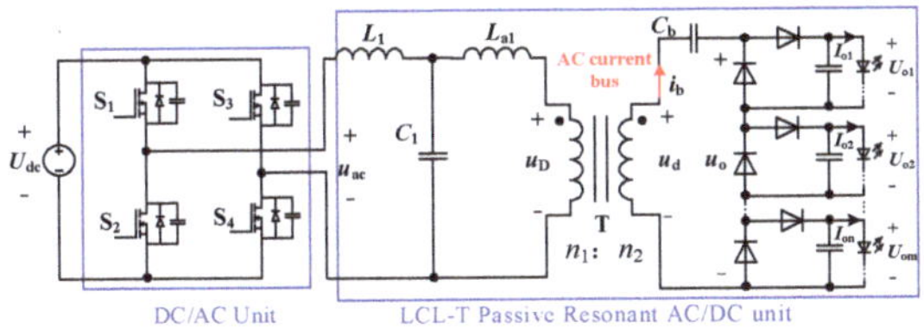

Figure 2. The proposed constant-current multi-output LED driver, based on a high-frequency AC bus.

Compared to [21], there are significantly fewer resonant networks, and the circuit structure is significantly simpler. Each string is connected in series before the diode rectifier, so only one output string needs to be sampled and closed-loop controlled. Other output currents can be automatically kept constant since all the strings are driven by the same AC bus. Obviously, high current accuracy can be achieved for multiple LED strings within the change of input and output.

However, for multi-output resonant converters, their stability is of crucial importance. Due to the nonlinear dynamic characteristics of the resonant converter, it is necessary to conduct stability analyses under various working conditions to ensure its stable operation under all conditions. If a linearized system is required, small-signal modeling is indispensable. This method is conducive to the rapid design of an ideal feedback controller [22]. For modern lighting systems, well-designed controllers must achieve high stability, rapid current regulation, provide fast transient response, and—at the same time—prevent current overtones that may damage LED [23]

This paper is organized as follows. Section 2 introduces the proposed DC-fed multi-channel constant-current LED driver. Section 3 elaborates on the parameter design process and provides a comparative stability analysis for the optimization of the PI controller. Section 4 details the implementation of the DC/DC control circuitry and verifies its functionality through experimental results. Finally, Section 5 concludes the paper.

2. The Proposed LED Driver

The topological structure proposed in this paper is an advancement upon and optimization of existing research. The modular topological structure proposed in reference [21] lays the key foundation for this study. This scheme utilizes the constant-current characteristics of the LCL-T network to drive the LED loads in each channel by connecting multiple LCL-T passive resonant rectifier units in parallel to the high-frequency AC voltage bus. This solution provides ideas for further optimization. Specifically, in this solution, each output channel is equipped with a dedicated LCL-T resonator. Therefore, this solution will inevitably increase the total number of components, cost, and physical volume, thereby limiting the power density.

To address the limitations of the aforesaid schemes, this letter presents a topology with a higher degree of integration, the core principle of which is to consolidate multiple resonant units into a centralized resonant network. This architecture utilizes the constant-current characteristics of an LCL-T network to establish a high-frequency AC bus, to which the rectifier bridges of each output channel are connected in series. This series configuration ensures that the output current of each channel is identical, thereby achieving inherent high-precision current balancing. Simultaneously, this approach significantly reduces the component count of passive elements. The final proposed scheme, derived from the analysis of this topology, is depicted in Figure 2.

The DC/AC inverter and the LCL-T passive resonant rectifier are the two components of the proposed DC-provided multi-output constant-current LED driver shown in Figure 2.

One LCL-T passive resonant network comes after the DC/AC inverter. After isolation by a transformer, a high-frequency AC bus is formed. Each output string is connected in series with the AC bus through the diode rectifier. Thus, multiple constant-current outputs can be realized with high current accuracy.

Consisting of switches S_1 through S_4, the DC/AC inverter transforms DC electricity into high-frequency AC square voltage. Additionally, Figure 1 shows the output voltage of the first PFC stage, which is the input DC voltage U_{dc}. An isolation transformer, T, a diode rectifier linked in series with the LED loads, a DC blocking capacitor, C_b, and resonant networks L_1, C_1, and L_{a1} make up the LCL-T passive resonant rectifier.

As illustrated in Figure 3, the converter operates in four primary modes within a single switching period, with the current paths highlighted in red and with inactive switching devices marked in gray. By employing phase-shift control, the full-bridge inverter converts the input DC voltage, U_{dc}, into a high-frequency square-wave voltage, u_{ac}.

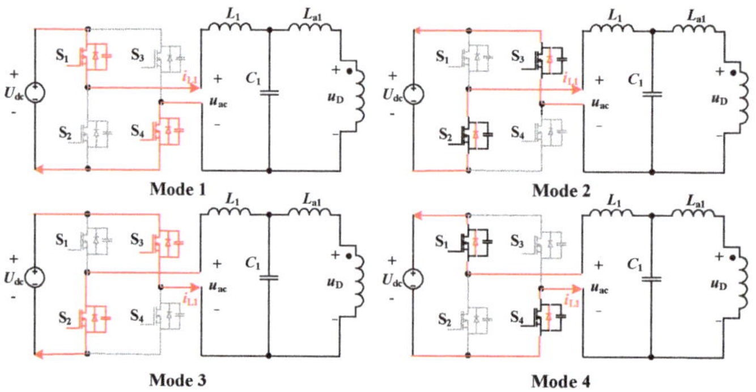

Figure 3. The four operational modes and corresponding current paths of the full-bridge inverter stage.

The operational modes are analyzed as follows:

In Mode 1, switches S_1 and S_4 are in the ON state. The DC power source forms a loop through S_1, the LCL-T resonant network, and S_4, transferring energy to the network.

In Mode 2, all the switches are turned OFF. The energy stored in the resonant inductor is released through the freewheeling diodes of S_2 and S_3, allowing the current to continue circulating.

In Mode 3, switches S_2 and S_3 are in the ON state. A reverse DC voltage is now applied to the input of the resonant network, and current flows through S_3, the resonant network, and S_2.

In Mode 4, S_2 and S_3 are turned OFF, and all the switches are once again in the OFF state. The energy in the resonant inductor is released through the freewheeling diodes of S_1 and S_4. This completes one full operational cycle.

Figure 4a displays the analogous circuit for the AC side. Phase-shifting full-bridge PWM control is applied to switches S_1–S_4. S_1 and S_2, and S_3 and S_4 are turned ON and OFF complementarily, respectively. S_1 and S_3 have the same duty cycle, and their phase difference is 180°. Thus, as seen in Figure 4b, the full-bridge DC/AC inverter's output voltage is a high-frequency AC square voltage, u_{ac}. The square voltage, u_o, serves as the diode rectifier's input voltage. The amplitude of u_o is equal to the sum of the LED load voltages when the current, i_b, is larger than 0. The amplitude of u_o equals 0 when the current, i_b, is smaller than 0.

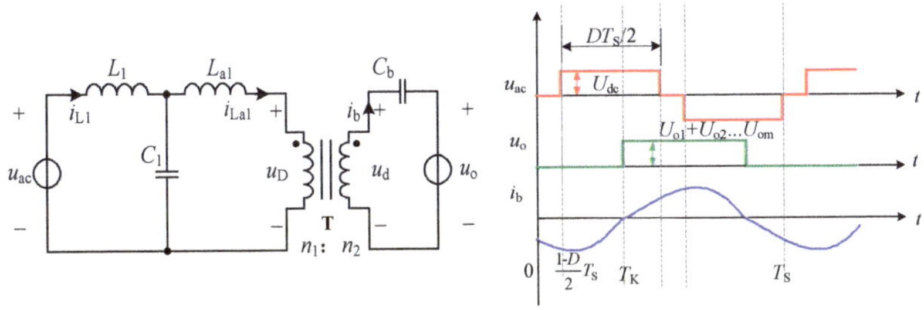

(**a**) The AC side equivalent circuit

(**b**) The main voltage and current waveforms of the proposed LED driver

Figure 4. The AC side equivalent circuit and the main voltage and current waveforms of the proposed LED driver.

As described in Figure 4b, the LCL resonant network, u_{ac}, input voltage can be written as:

$$u_{ac} = \begin{cases} 0 & \left(0 < t \leq \frac{1-D}{2}T_S\right) \\ U_{dc} & \left(\frac{1-D}{2}T_S < t \leq \frac{T_S}{2}\right) \\ 0 & \left(\frac{T_S}{2} < t \leq \frac{2-D}{2}T_S\right) \\ -U_{dc} & \left(\frac{2-D}{2}T_S < t \leq T_S\right) \end{cases} \tag{1}$$

where D is the switching device's conduction duty cycle, and T_S is its switching period.

According to the above analysis of the square voltage, u_o, it can be expressed as follows in the first switching cycle.

$$u_o = \begin{cases} 0 \ (0 < t \leq T_K) \\ U_o \ \left(T_K < t \leq T_K + \frac{T_S}{2}\right) \\ 0 \left(T_K + \frac{T_S}{2} < t \leq T_S\right) \end{cases} \tag{2}$$

where $U_o = U_{o1} + U_{o2} + ... + U_{om}$, m is the number of output channels, and T_K is the starting delay time of the output voltage, u_o, square wave.

Both u_{ac} and u_o are AC square voltages, so their Fourier series expansion can be easily obtained as:

$$u_{ac} = \sum_{h=1,3,5,7...} U_{achm} \sin(h\omega_S t + \varphi_h) \tag{3}$$

where

$$\begin{aligned} \varphi_h &= \tan^{-1}\frac{\sin(hD\pi)}{\cos(hD\pi)-1} \\ U_{achm} &= \sqrt{a_h{}^2 + b_h{}^2} \\ a_h &= [\sin(D-1)h\pi - \sin(Dh\pi)]\frac{U_{dc}}{h\pi} \\ b_h &= [1 - \cos(h\pi) - \cos(Dh\pi) + \cos(D-1)h\pi]\frac{U_{dc}}{h\pi} \end{aligned}$$

And the voltage, u_o, can be expressed as

$$u_o = \frac{U_o}{2} + \sum_{h=1,3,5,7...} U_{ohm} \sin[h\omega_S(t - T_K)] \tag{4}$$

where

$$U_{ohm} = \frac{2U_o}{h\pi}, U_o = U_{o1} + U_{o2} + \ldots + U_{om}$$

DC-blocking capacitor C_b filters out the DC component of the high-frequency square voltage, u_o, which is $U_o/2$. Therefore, the LCL resonant network output voltage, u_D, may be written as

$$u_D = \sum_{h=1,3,5,7\ldots} U_{Dhm} \sin[h\omega_S(t - T_K)] \tag{5}$$

where

$$U_{Dhm} = \frac{2NU_o}{h\pi}$$

Additionally, N is the transformer T's turns ratio, which is equivalent to n_1/n_2.

The output currents of each LED string are identical since they are linked in series via the diode rectifier, which means

$$I_{o1} = I_{o2} = \ldots = I_{om} \tag{6}$$

Figure 2 shows that the precise output current, $I_{o1\text{-act}}$, is equivalent to half of the average value of the current, i_b, during the positive half-switching phase. So, if the currents i_b or i_{La1} can be obtained quantitatively, the exact solution of the output current $I_{o1\text{-exa}}$ can be obtained, which is derived as

$$I_{o1-exa} = \frac{1}{2\pi} \int_{\omega_S T_K}^{\omega_S T_K+\pi} (i_b) d\omega_S t = \frac{N}{2\pi} \int_{\omega_S T_K}^{\omega_S T_K+\pi} (i_{La1}) d\omega_S t \tag{7}$$

Next, the characteristics of the output current are analyzed considering the fundamental and higher harmonics, respectively.

2.1. Only Considering the Fundamental Harmonic

The ratio of inductors L_1 and L_{a1}, the resonant angular frequency, ω_o, and its normalized angular frequency, ω_n, are defined as

$$\omega_o = \frac{1}{\sqrt{L_1 C_1}},\ \omega_n = \frac{\omega_S}{\omega_o},\ \omega_S = 2\pi f_S,\ \gamma = \frac{L_1}{L_{a1}} \tag{8}$$

where f_S is the switching frequency, and ω_S represents its angular frequency.

The LCL-T rectifier functions as an ideal current source if the resonant angular frequency is equivalent to the switching frequency, or $\omega_n = 1$, and only takes into account the fundamental harmonic in voltage u_{ac} and u_D [21]. Accordingly, as examined in [21], the output current, I_{o1}, is independent of the output voltage, and it can be expressed as

$$I_{o1} = \frac{N}{\pi} \frac{U_{ac1m}}{\omega_S L_1} \tag{9}$$

The LCL-T resonant network parameters may be found using Equations (7) and (8). If U_{ac1m}, N, and I_{o1} are known, the inductor, L_1, can be calculated from (8), and the capacitor C_1 can be deduced from (7).

The determination of the inductor, L_{a1}, is as follows. When γ is equal to 1, the fundamental component of the resonant network input current, i_{L1}, is in phase with the fundamental component of the voltage, u_{ac}, as concluded in [21]. The resonant network should be slightly inductive to achieve the zero-voltage switching conditions for switches, so γ needs to be greater than 1. As γ is taken as 2 here, it is also verified that it is a reasonable value in Section 3. Next, L_{a1} can be calculated from (8) once L_1 and γ are determined.

2.2. Considering Higher Harmonics

Figure 5 below represents the analogous circuit of the LCL-T resonant network seen in Figure 4a. There are many higher harmonics in the current i_{La1} because the voltages, u_{ac} and u_D, of the LCL-T network correspond to the AC square voltages from (3) and (5), and these include considerably higher harmonics. It is crucial to carefully examine how higher harmonics affect the precision of the output currents, since (7) will have an impact on the accuracy of the precise output current, $I_{o1\text{-}exa}$.

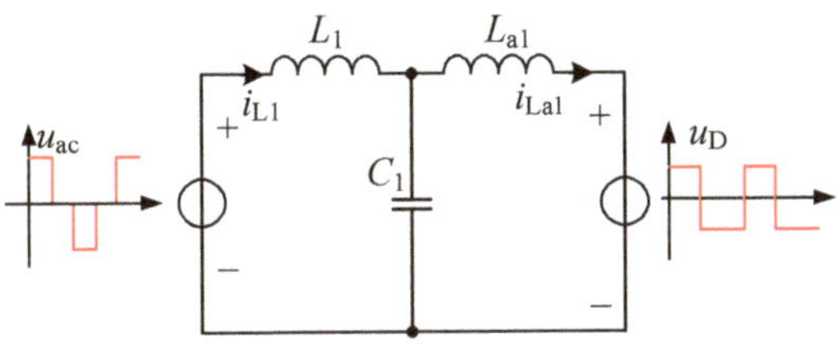

Figure 5. The simplified equivalent circuit of the LCL-T resonant network.

In Figure 5, i_{La1} can be solved when the voltage sources, u_{ac} and u_D, work separately. Then, the final current, i_{La1}, can be obtained according to the superposition theorem. The expressions of u_{ac} and u_D are shown in (3) and (5), and the phase of the AC voltage, u_D, is determined by T_K, which is solved as follows.

The AC voltage, u_D, crosses zero when $t = T_K$, and the inductor current, i_{La1}, also crosses zero at this time, that is, $i_{La1}(T_K) = 0$. In order to determine the connections between T_K, U_{dc}, N, and U_o using the frequency domain analysis approach, Equation (10) is obtained from [21]. The output current, $I_{o1\text{-}exa}$, can thus be precisely solved as (11). Finding the precise current solution and examining the parameters needed to satisfy the output current's accuracy criteria are the primary uses of Equation (11).

$$-\sum_{h=1,3,5,7\ldots}^{\infty}\frac{\gamma U_{achm}\cos(h\omega_S T_K+\varphi_h)}{h[\gamma-(h^2-1)]}+\sum_{h=1,3,5,7\ldots}^{\infty}\frac{\gamma(1-h^2)U_{Dhm}}{h[\gamma-(h^2-1)]}=0 \quad (10)$$

After determining U_{dc}, N, D, and U_o, T_K may be solved using mathematical programs like MATLABR2024a, as stated in [21]. Then, the precise solution of the output current, $I_{o1\text{-}exa}$, is obtained as follows:

$$\begin{aligned} I_{o1-exa} &= \frac{N}{2\pi}\int_{\omega_S T_K}^{\omega_S T_K+\pi}(i_{La1})d\omega_S t \\ &= \frac{N}{2\pi}\left\{\sum_{h=1,3,5,7\ldots}^{\infty}\frac{2\gamma U_{achm}}{h^2[\gamma-(h^2-1)]\omega_S L_1}\sin(h\omega_S T_K+\varphi_h)\right\} \end{aligned} \quad (11)$$

As indicated by (10) and (11), $I_{o1\text{-}exa}$ is a function of U_{dc}, U_o, N, D, and L_1. Therefore, the subsequent step is to determine the parameters, N and L_1. The architecture of the LCL-T passive resonant rectifier provides the conditions necessary to achieve zero-voltage switching (ZVS), the realization of which is ensured by the proper selection of the parameter γ.

To validate the preceding theoretical analysis, a corresponding simulation model was constructed. As depicted in Figure 6, the simulation results clearly demonstrate that the input voltage, u_{ac}, (red trace) is a square wave with an amplitude of ±400 V, while the output voltage, u_D, (blue trace) is also a square wave that lags u_{ac}. This outcome is in good agreement with the theoretical analysis for the equivalent voltage at the output of the LCL-T resonant network, thus verifying the correctness of the theory presented in this section.

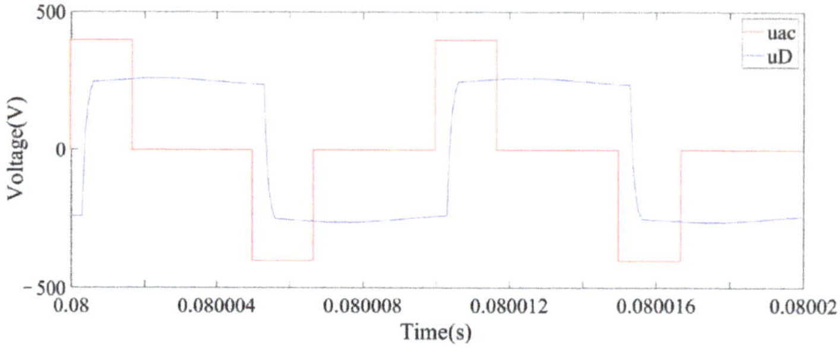

Figure 6. Voltage on both sides of the LCL-T resonant network in the simulation model.

The primary function of the LCL-T resonant network is to perform waveform shaping. Through the analysis and calculation of inductor L_1 and capacitor C_1, its resonant frequency is set to the switching frequency. The network exhibits low impedance to the fundamental component of the square-wave voltage, u_{ac}, while presenting high impedance to higher-order harmonics. Furthermore, the presence of the inductor, L_{a1}, effectively attenuates high-order harmonics. The DC-blocking capacitor, C_b, connected in series with the transformer secondary, serves to filter out any DC bias resulting from various factors, thereby ensuring the stable operation of the entire resonant network. In summary, owing to the filtering effect of the LCL-T network and the DC-blocking protection afforded by C_b, the system successfully converts the square-wave voltage into a smooth, high-quality AC.

3. Parameter Design

The stable state model of the whole system, taking into account only the fundamental harmonic, may be obtained using Equations (3) and (9):

$$I_o = \frac{4NU_{dc}}{\pi^2 \omega_S L_1} \sin\left(D\frac{\pi}{2}\right) \tag{12}$$

By linearizing (12) around the steady-state operating point (D, I_o), the influence of a duty cycle perturbation on the output current can be obtained. This is defined as the system's DC gain, K_{dc}:

$$K_{dc} = \frac{\partial I_o}{\partial D} = \frac{2NU_{dc}}{\pi \omega_S L_1} \cos\left(D\frac{\pi}{2}\right) \tag{13}$$

Through the output filtering stage, the system's low-frequency dynamic properties are established. The output filter for this topology may be roughly represented via the first-order RC low-pass filter, and the transfer function, $H(s)$, is as follows:

$$H(s) = \frac{1}{1 + sR_{LED}C_o} \tag{14}$$

where C_o is the output filter capacitance, and R_{LED} is the equivalent resistance of the LED load.

By combining the DC gain and low-frequency dynamic characteristics obtained from the above analysis, the small-signal transfer function, $G(s)$, can be obtained as follows:

$$G(s) = \frac{\hat{i}_o(s)}{\hat{d}(s)} = K_{dc} \cdot H(s) = \frac{\frac{2NU_{dc}}{\pi \omega_S L_1} \cos\left(D\frac{\pi}{2}\right)}{1 + sR_{LED}C_o} \tag{15}$$

Equation (15) was derived using the fundamental harmonic approximation (FHA) method, which represents the open-loop small-signal transfer function of the entire system. This model abstracts the controlled object as a first-order system with variable gain K_{dc}, primarily used to analyze the low-frequency dynamic characteristics of the system and lay the groundwork for subsequent feedback controllers.

To verify the correctness and accuracy of this theoretical model, a simulation model of the proposed LED driver was constructed in MATLAB/Simulink.

Since the simulation's circuit parameters act as a direct reference for the practical prototype, they are extremely important. The simulation was configured with a DC input voltage, U_{dc}, of 400 V and a switching frequency, f_s, of 100 kHz. To reduce the physical volume of the resonant network, the transformer turns ratio, N, was set to 2 after detailed analysis. For a target constant output current, I_o, of 0.7 A, with the normalized angular frequency, ω_n, set to 1 and the inductor ratio, γ, preset to 2, Equations (8) and (9) were used to compute the LCL-T resonance network's constants. This yielded the following values: L_1 = 364.5 μH, L_{a1} = 182.25 μH, and C_1 = 6.96 nF.

To validate the accuracy of the theoretical model, the transfer function, $G(s)$, derived from the theoretical analysis, is compared against $Gf(s)$ obtained from the simulation model. Figure 7 presents a comparison of the theoretical and simulated Bode plots. As illustrated in the figure, the Bode plot derived from the theoretical calculation is in good agreement with the frequency-sweep simulation results, thereby validating the correctness of the established mathematical model in the low-frequency domain.

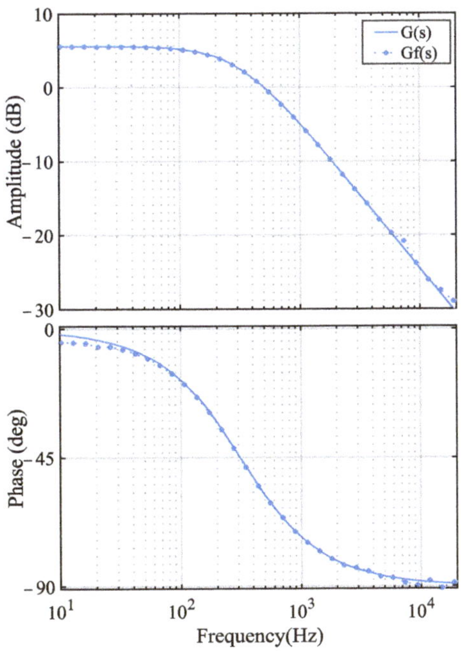

Figure 7. The Bode plots of $G(s)$ and $Gf(s)$.

Following the simulation verification, the subsequent step involves preparing for the experimental implementation. To design a closed-loop controller capable of fast and precise output current regulation, a thorough understanding of the system's model characteristics is essential. The small-signal function of transfer $G(s)$ from the obligation cycle, $d(s)$, to the output current, $i_o(s)$, demonstrates the traits of a conventional first-order low-pass filter, as per Equation (15):

$$G(s) = \frac{\hat{i}_o(s)}{\hat{d}(s)} = K_{dc} \cdot H(s) = \frac{K_{dc}}{1 + s/\omega_p}$$

where the pole $\omega_p = 1/(R_{LED} \cdot C_o)$.

By substituting the relevant circuit parameters, it can be calculated that the system pole, f_p, is 19.5 Hz. The system's dynamic reaction time is quite sluggish because of the pole's relatively low frequency, which prevents it from meeting the necessary requirements for quick adjustment and quick startup.

Additionally, the duty cycle, D, change affects the DC gain, K_{dc}. When duty cycle D is zero, the worst-case situation arises, and K_{dc} achieves its maximum value, represented by K_{dc_max}:

$$K_{dc\_max} = \frac{2NU_{dc}}{\pi\omega_S L_1}$$

Following the above theoretical analysis, the designed PI controller needs to meet four requirements. To enhance the system's dynamic reaction speed, the bandwidth must first be increased (f_c = 10 kHz). Secondly, it is also necessary to ensure that the system has a sufficient stability margin to resist parameter changes and external disturbances. Thirdly, the compensated system must eliminate the steady-state error, which is the basic requirement for maintaining a constant output current. Fourth, the system also needs to reduce overshoot to protect the LED components and extend their service life.

After analysis, by appropriately adjusting the PI parameters, the proportional–integral controller can meet the above four requirements.

In this section, three distinct PI controller design methods are compared, all based on the worst-case gain scenario. These methods are the zero-pole cancellation method, the phase margin tuning method, and the pole placement method. A comparative analysis of various performance metrics will be conducted to determine the most suitable design approach for the experiment.

To provide a clear visual comparison, the open-loop transfer functions resulting from each of the three methods are plotted using MATLABR2024a, as shown in Figure 8. In the figure, the uncompensated system is represented by the curve, $G(s)$, while the systems compensated using methods one, two, and three are represented by $L_1(s)$, $L_2(s)$, and $L_3(s)$, respectively. Meanwhile, after calculation, various data of the open-loop curves in the figure are supplemented in Table 1 for convenient comparison and analysis.

Table 1. Comparative analysis of PI controller design methods.

System	K_p	K_i	f_c (kHz)	PM (deg)	Overshoot (%)	Settling Time (μs)
G(s)			0.04	115.6	0.0	9630.6
Method 1	221.4662	27,130.49	10.0	90.0	0.0	62.3
Method 2	191.5795	6,981,061.78	10.0	60.0	24.3	149.8
Method 3	221.4666	6782.64	10.0	90.1	0.0	63.4

The plant model, based on the system parameters, is:

$$G(s) = K_{dc\_max}/(1 + s/\omega_p) = 2.9135/(1 + s/122.51)$$

The resulting open-loop transfer functions corresponding to the curves $L_1(s)$, $L_2(s)$, and $L_3(s)$ are as follows:

For $L_1(s)$: The controller $G_{c1}(s)$ precisely sets its zero point at the pole of the controlled object, thereby achieving zero-pole cancellation and simplifying the system to a pure integrator form.

$$L_1(s) = G_{c1}(s)G(s) = \left[K_{p1}(s+\omega_{z1})/s\right]\left[K_{dc\_max}/(1+s/\omega_p)\right]$$
$$L_1(s) = (K_{p1}K_{dc\_max}\omega_p)/s \approx 357.0/s$$

For $L_2(s)$: By reasonably designing the parameters of controller $G_{c2}(s)$, the system can achieve the specified phase margin. The complete open-loop transfer function is:

$$\begin{aligned} L_2(s) &= G_{c2}(s)G(s) = \left[K_{p2}(s+\omega_{z2})/s\right]\left[K_{dc\_max}\ (1+s/\omega_p)\right] \\ L_2(s) &= [(191.58s+6981062)/s][2.9135/(1+s/122.51)] \\ &= (557.9s+20340000)/(0.00816s^2+s) \end{aligned}$$

For $L_3(s)$: The controller, $G_{c3}(s)$, uses the pole placement method for its zero location. The complete open-loop transfer function is:

$$\begin{aligned} L_3(s) &= G_{c3}(s)G(s) = \left[K_{p3}(s+\omega_{z3})/s\right]\left[K_{dc\_max}/(1+s/\omega_p)\right] \\ L_3(s) &= [(221.47s+6783)/s][2.9135/(1+s/122.51)] \\ &= (645.2s+19760)/(0.00816s^2+s) \end{aligned}$$

The transfer functions of $L_1(s)$, $L_2(s)$, and $L_3(s)$ are obtained through the above theoretical calculations. Based on their transfer functions, the corresponding Bode plots can be drawn.

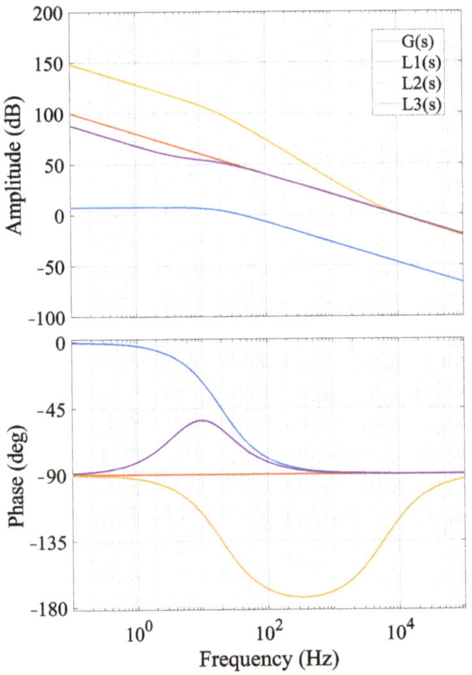

Figure 8. The Bode plots of $G(s)$, $L_1(s)$, $L_2(s)$, and $L_3(s)$.

By analyzing the performance metrics of each curve in Table 1 and the frequency response characteristics shown in Figure 8, the zero-pole cancellation method (Method 1) is the most suitable compensation strategy for this application. Its principle lies in aligning the controller's zero with the main pole of the controlled object, achieving precise cancellation functionally, thereby transforming the open-loop transfer function into an approximate ideal integrator. This method exhibits excellent robustness, characterized by a 90-degree phase margin and zero overshoot characteristics, which protect the LED load from overcurrent hazards. Given the significant advantages of the zero-pole cancellation method, this method is also adopted in the subsequent experimental verification to design the PI controller.

4. Experimental Verification and Analysis

The preceding sections explain the operating principle of the converter, analyze and derive the mathematical model and small-signal model of the system, and design the most suitable PI controller for this paper based on theoretical analysis.

Based on the above system analysis and controller design, this section constructs a complete closed-loop control system, as shown in Figure 9. In order to verify the validity of the aforementioned analysis and parameter design, a five-channel experimental prototype with a rated power of 200 W was constructed, and its detailed specifications are the same as those listed in Table 2. The corresponding circuit components are also listed in the same table. The control strategy comprises the following: firstly, the current of a single output channel is sampled; secondly, this is compared with a preset reference signal to generate an error signal; thirdly, the error signal is processed by a PI compensator to output a modulating signal; and finally, the modulating signal is compared with a triangular carrier to generate PWM signals to drive the four power switches. The series connection of the output channels via rectifier circuits enables the system to achieve effective closed-loop control of all output currents. As illustrated in Figure 9, the block diagram depicts the closed-loop control system, with its core control unit implemented using a TMS320F28335 digital signal processor. Figure 10 is a photograph of the experimental prototype.

Table 2. Experimental parameters of the prototype.

Symbol	Value
K_p	221
K_i	27,130
f_S	100 kHz
N	2
m	5
L_1	350 μH
L_{a1}	175 μH
C_1	7 nF
C_b	1 uF
S_1–S_4	IPP60R099C7
Control board	TMS320F28335

In the control circuitry of the experimental prototype, the isolated gate driver is a critical component. The isolated driver circuit employed in this study is depicted in Figure 11, wherein pwm3 is the PWM input signal. The gate driver IC, IC1 (NSi6602B-DSWR), converts the logic-level pulse (PWM3) from the controller into a high-frequency square-wave drive signal. This conversion is achieved by its internal push–pull output stage, which alternately applies high and low levels across the PWMC and GNDC output terminals, connected respectively to the gate and source of the power MOSFET. The switching frequency of the MOSFET is set to 100 kHz.

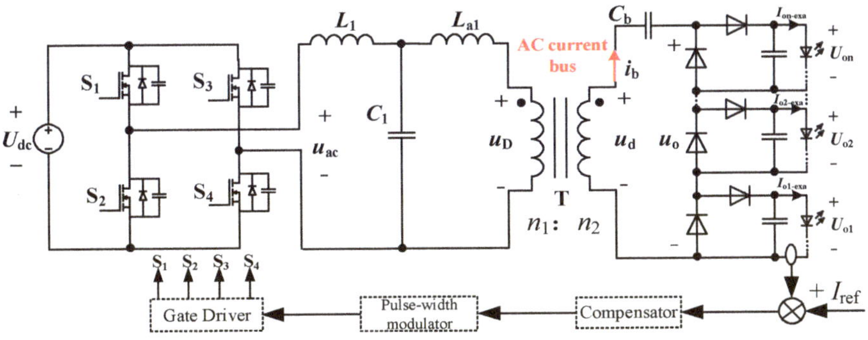

Figure 9. The suggested LED driver prototype.

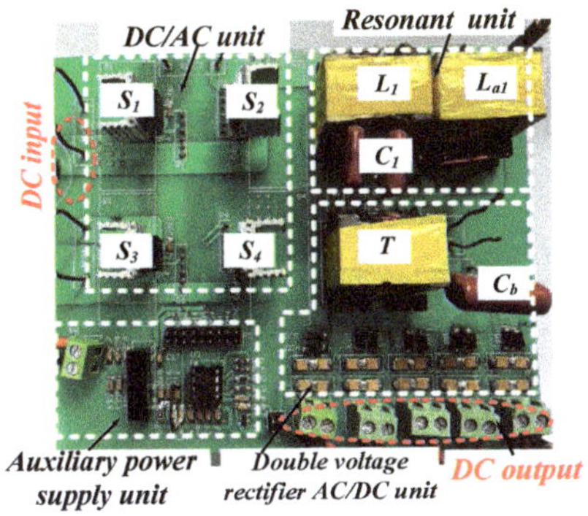

Figure 10. The prototype of the proposed LED driver.

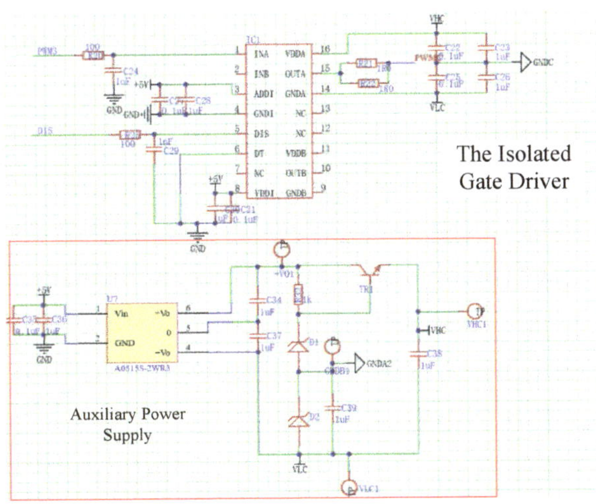

Figure 11. Isolated gate driver circuit diagram.

4.1. Experimental Waveform Analysis

This section presents the performance verification results of the experimental prototype. As shown in Figure 12a–c, the waveform diagrams illustrate the steady-state output waveforms under rated power and different input voltage conditions. The figure shows the experimental waveforms under operating conditions with input voltages, U_{dc}, of 380 V, 400 V, and 420 V. In each subfigure, the waveforms from top to bottom are the drain–source voltage of switch S_1 (u_{ds1}), the channel output current ($I_{o1\text{-}exa}$), and the final output voltage (U_{o1}). The experimental data indicate that the output current, Io1-exa, remains at a constant level of 700 mA within the input voltage range of 380 V to 420 V. This experimental result validates the effectiveness of the closed-loop control strategy and meets the system design specifications for high-precision output current requirements.

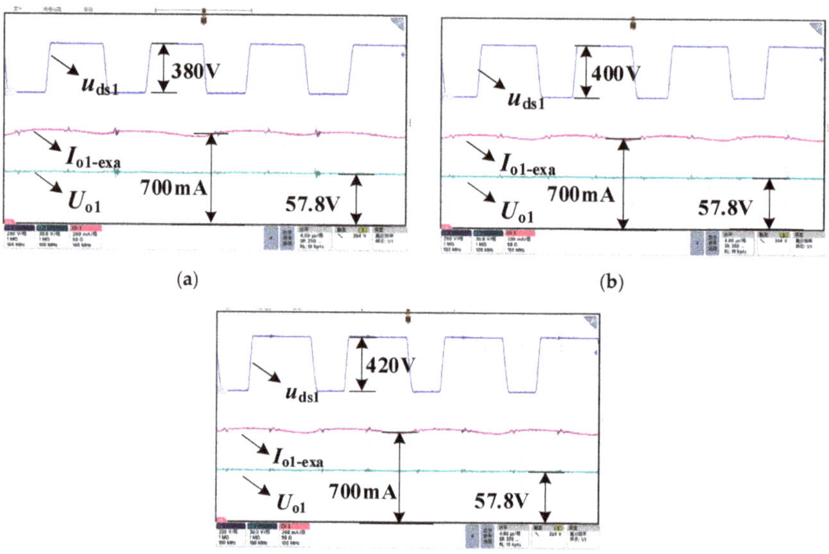

(a) (b) (c)

Figure 12. The steady-state output waveforms with input voltage changes. (**a**) The DC bus voltage is 380 V; (**b**) The DC bus voltage is 400 V; (**c**) The DC bus voltage is 420 V.

The objective of this study is to verify the zero-voltage switching (ZVS) conditions described in the theoretical analysis of the first part by employing switch transistors S_1 and S_3 as research objects and utilizing experimental characteristic analysis methods. As demonstrated in Figure 13a, under operating conditions of input voltage U_{dc} = 400 V and output power P = 200 W, the drain–source voltage (u_{ds1}) and channel current (i_{s1}) waveforms of switch transistor S_1 were measured. As illustrated in Figure 13b, the measured values correspond to the drain–source voltage (u_{ds3}) and channel current (i_{s3}) of switch transistor S_3. The experimental data suggest that the currents, i_{s1} and i_{s3}, are both negative prior to each transistor switching operation. This negative state has been proven to cause the anti-parallel diodes to enter a forced conduction state. This finding confirms that the switching transistors, S_1 and S_3, have successfully achieved zero-voltage switching (ZVS) operation.

Figure 14 shows transient waveforms when the input voltage, U_{dc}, changes at rated power 200 W. Figure 15 shows transient waveforms when the output voltage, U_{o1}, changes. Taking the first and second output channels as an example, the waveforms in Figure 14 are the input voltage, U_{dc}, and the output currents, $I_{o1\text{-}exa}$ and $I_{o2\text{-}exa}$. As illustrated in Figure 15, the dynamic waveforms of the output voltage, U_{o1}, and the output currents, $I_{o1\text{-}exa}$ and $I_{o2\text{-}exa}$, are also shown. The experiment demonstrates that, in conditions of load disturbance, the output current can rapidly revert to the set value within a brief timeframe. This outcome validates the efficacy of closed-loop control for all output channels and substantiates the accuracy of the output current, which aligns with the design specifications across the entire operational range.

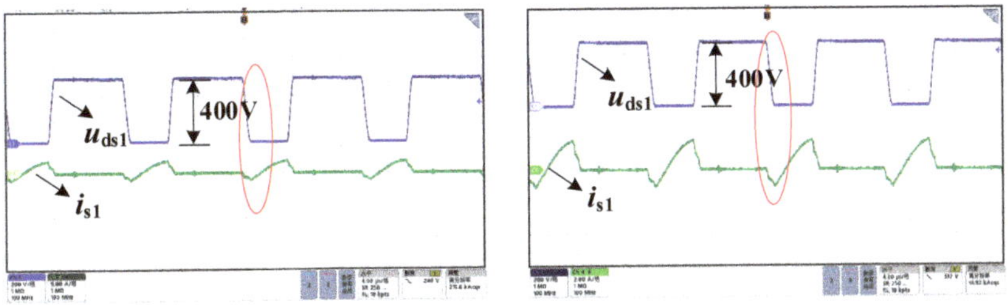

(**a**) (**b**)

Figure 13. The measured drain–source voltage of S_1 and S_3 and the corresponding currents. (**a**) Switching device S_1; (**b**) Switching device S_3.

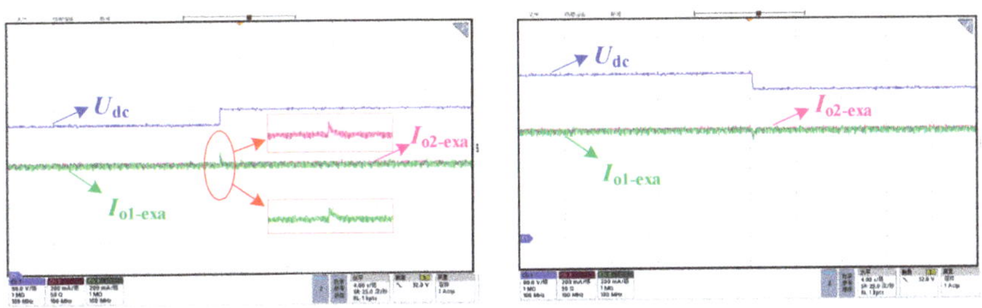

(**a**) (**b**)

Figure 14. The transient waveforms with input voltage changes. (**a**) The DC bus voltage ranges from 380 V to 420 V; (**b**) The DC bus voltage ranges from 420 V to 380 V.

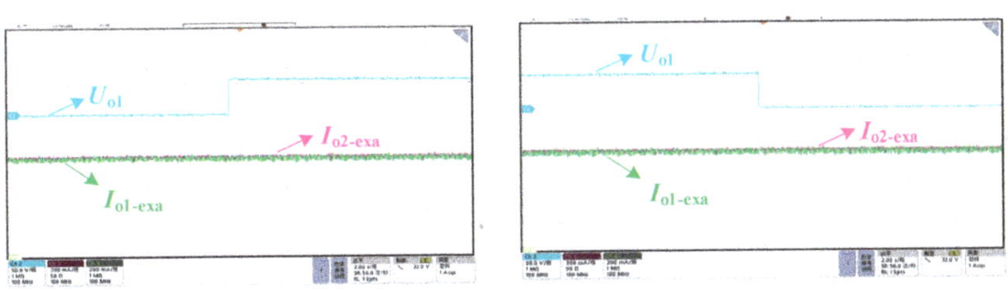

(**a**) (**b**)

Figure 15. The transient waveforms with output voltage changes. (**a**) The output voltage of the first path ranges from 0 V to 57 V; (**b**) The output voltage of the first path ranges from 57 V to 0 V.

4.2. Loss Analysis and Theoretical Efficiency Estimation

By thoroughly analyzing the wear and tear on each component, we have provided a theoretical basis for proving the high efficiency of the system. Figure 16 illustrates the loss model of the converter, which incorporates the primary parasitic components inherent in the circuit elements. These components include the on-resistance ($r_{DS}(on)$) of the MOSFET, the forward voltage drop (V_F) and on-resistance (r_F) of the diode, the parasitic resistance

(r_{L1} and r_{La1}) of inductors L_1 and L_{a1}, the parasitic resistance of capacitors C_1 and C_b (r_{C1} and r_{Cb}), and the primary and secondary winding resistances of the transformer (r_{pri} and r_{sec}), which are the fundamental sources of power loss. Based on this model, the power loss calculations for each component are as follows.

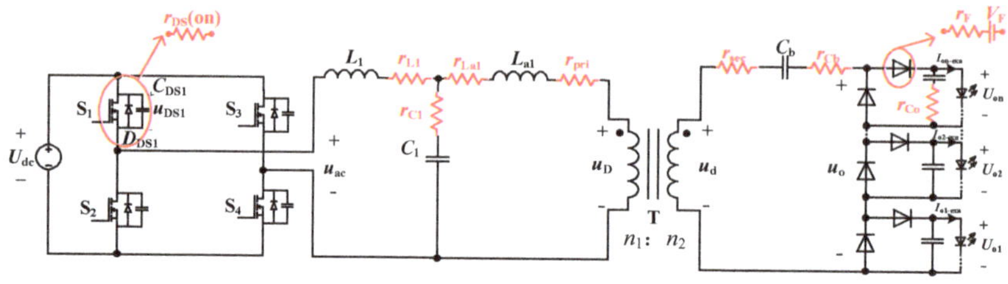

Figure 16. Circuit diagram of loss model with parasitic parameters.

The total conduction loss of the four switches can be calculated based on the effective value, $I_{L1,rms}$, of the current, i_{L1}, flowing through the AC side of the inverter:

$$P_{rDS\_total} = 2 \times I^2_{L1,rms} \times r_{DS}(on) \tag{16}$$

This topology achieves zero-voltage switching (ZVS) through phase shift control, so switching losses can be ignored. Losses mainly originate from the switching process, and the switching loss of a single switch can be expressed as:

$$P_{sw} = f_S \cdot C_{DS1} \cdot U^2_{dc} \tag{17}$$

where f_S is the switching frequency, and C_{DS1} is the output capacitance of the MOSFET.

Therefore, the total MOSFET loss is:

$$P_{Switch} = P_{rDS\_total} + 4P_{sw} \tag{18}$$

Diode conduction losses are caused by both the forward voltage drop, V_F, and the equivalent dynamic resistance, r_F. The diode loss for a single channel is:

$$P_{diode,ch_n} = V_F I_{on-exa} + I^2_{diode,rms} r_F \tag{19}$$

where $I_{diode,rms}$ is the effective value of the current flowing through a single rectifier diode.

$$P_{Diode\_total} = \sum_{n=1}^{m} P_{diode,ch_n} \tag{20}$$

The loss of inductance is mainly caused by copper loss generated by parasitic resistance, r_L. For inductances L_1 and L_{a1} in this circuit, the total loss is:

$$P_{rL\_total} = r_{L1} \cdot I^2_{L1,rms} + r_{La1} \cdot I^2_{La1,rms} \tag{21}$$

where $I_{L1,rms}$ and $I_{La1,rms}$ are the effective values of the alternating current flowing through the corresponding inductors.

Capacitor loss is caused by its parasitic resistance, r_C. The total capacitor loss is:

$$P_{RC\_total} = r_{C1} \cdot I^2_{C1,rms} + r_{Cb} \cdot I^2_{Cb,rms} + \sum_{n=1}^{m} (r_{Co} \cdot I^2_{Co,n,rms}) \tag{22}$$

where $I_{C1,rms}$, $I_{Cb,rms}$, and $I_{Co,n,rms}$ are the effective values of the alternating current flowing through the corresponding capacitors.

Transformer losses include copper losses and iron losses:

$$P_{T\_total} = (r_{pri} I_{pri,rms}^2 + r_{sec} I_{sec,rms}^2) + P_{core,T} \tag{23}$$

where $I_{pri,rms}$ and $I_{sec,rms}$ are the effective values of the primary and secondary currents, respectively, and $P_{core,T}$ is the magnetic loss.

Adding up all of the above loss components to obtain the total theoretical loss of the converter results in:

$$P_{loss} = P_{SW\_total} + P_{Diode\_total} + P_{L\_total} + P_{C\_total} + P_{T\_total} \tag{24}$$

To theoretically assess the efficiency of the proposed converter, the theoretical efficiency of the system can be estimated by the following equation:

$$\eta_{theory} = \frac{P_{out}}{P_{in}} = \frac{P_{out}}{P_{out} + P_{loss}} \tag{25}$$

This section provides a quantitative analysis of the primary power loss sources in the system, based on device datasheets, circuit simulations, and standard engineering estimation methods. Under rated operating conditions, the calculated theoretical total loss is approximately 9.45 W, corresponding to an efficiency of 95.5%. Among these losses, the rectifier diodes account for the largest portion, exceeding 58%, followed by the core loss of the magnetic components (approx. 21%) and the conduction loss of the MOSFETs (approx. 10%).

4.3. Experimental Efficiency Analysis

To validate the accuracy of the theoretical analysis, efficiency tests were conducted on the prototype. Figure 17 presents the measured efficiency curves under various input/output conditions. The experimental results indicate that the efficiency of the topology improves significantly as the output load increases, with a measured peak efficiency of 94.4% being achieved under rated operating conditions.

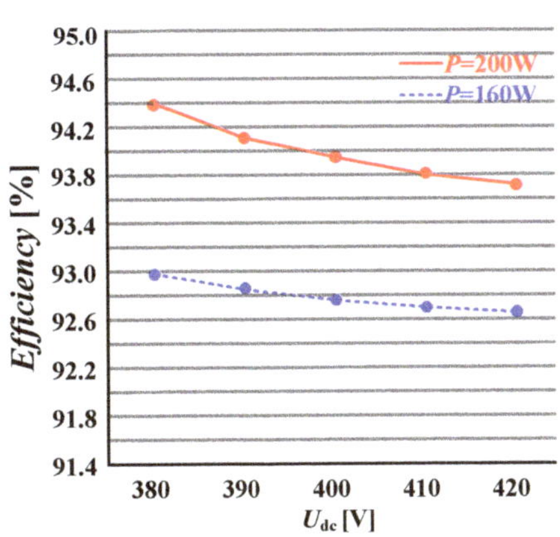

Figure 17. The dynamic waveforms with output voltage changes.

In summary, the experimental tests confirm the precise control capability of the apparatus over the output voltage and current. This measured value is in close agreement with the theoretical peak efficiency of 95.5%, with the minor discrepancy being primarily attributed to unmodeled parasitic parameters and measurement errors. This result further validates the correctness of the design methodology and its practical engineering value.

5. Conclusions

In this paper, the proposed solution is an LED driver with a multi-output configuration that operates on a high-frequency AC bus. Closed-loop regulation of all the series circuits can be achieved by sampling and controlling a single output series circuit. A theoretical analysis of the impact of higher-order harmonics on the constant-current output was conducted, and an exact solution for the output current was derived. After verifying the validity of the topological theory, various PI controller design methods were compared and analyzed, with the zero-pole elimination method ultimately selected. Finally, a five-channel 200 W experimental prototype was constructed, and the effectiveness of the theoretical analysis was validated through test results. The tests demonstrated that all the series circuits meet the high current accuracy requirements.

Author Contributions: Conceptualization, Q.H. and Z.T.; Methodology, Q.H., Z.T. and W.Z.; Software, Z.T.; Validation, Q.H., Z.T., W.Z. and K.Z.; Resources, Q.H.; Writing—original draft, Q.H., Z.T. and W.Z.; Writing—review & editing, Q.H., Z.T., W.Z. and K.Z.; Project administration, Q.H. and K.Z.; Funding acquisition, Q.H. All authors have read and agreed to the published version of the manuscript.

Funding: This work was supported by the National Natural Science Foundation of China (52307219) and the Wuhan Science and Technology Bureau Project (2023020201020307).

Data Availability Statement: Data is contained within the article.

Conflicts of Interest: The authors declare no conflict of interest.

References

1. Qu, Y.; Qiu, L. I2V2 Average Current Control for Modular LED Drivers. *IEEE Trans. Power Electron.* **2021**, *36*, 78–82. [CrossRef]
2. Huang, Y.; Ma, D.B. A Smart Silicon Carbide LED Driver IC With Integrated Dual-Level Condition-Monitoring Mechanism. *IEEE Trans. Power Electron.* **2024**, *39*, 6246–6255. [CrossRef]
3. Zhang, J.; Shen, H.; Du, X.; Chen, R. Condition Monitoring the Inhomogeneous Thermal Fatigue of Multichip IGBT Module Based on the Thermal Attenuation Coefficient. *IEEE Trans. Power Electron.* **2025**, *40*, 2114–2125. [CrossRef]
4. Zhang, J.; Du, X.; Qian, C.; Du, R.; Hu, X.; Tai, H.-M. Thermal Management of IGBT Module in the Wind Power Converter Based on the ROI. *IEEE Trans. Ind. Electron.* **2022**, *69*, 8513–8523. [CrossRef]
5. Li, H.; Li, S.; Xiao, W. Single-Phase LED Driver with Reduced Power Processing and Power Decoupling. *IEEE Trans. Power Electron.* **2021**, *36*, 4540–4548. [CrossRef]
6. Esteki, M.; Khajehoddin, S.A.; Safaee, A.; Li, Y. LED Systems Applications and LED Driver Topologies: A Review. *IEEE Access* **2023**, *11*, 38324–38358. [CrossRef]
7. Mukherjee, S.; Yousefzadeh, V.; Sepahvand, A.; Doshi, M.; Maksimović, D. A Two-Stage Automotive LED Driver With Multiple Outputs. *IEEE Trans. Power Electron.* **2021**, *36*, 14175–14186. [CrossRef]
8. Song, J.; Kim, M.; Park, H.; Jung, W.; Ahn, Y.; Nam, T.; Shin, Y.; Woo, Y.-J.; Lee, H.-M. A 96.3%-Efficiency Wide-Voltage-Range Three-Level Boost Converter with Loop-Free Self-Balancing and Dead-Zone PWM Control for Backlight LED Drivers. *IEEE Trans. Power Electron.* **2024**, *39*, 7064–7075. [CrossRef]
9. Wang, Y.; Xu, J.; Chen, Z.; Qin, F. Small Signal Modeling and Transient Cross-Regulation Analysis of Peak-Current-Mode-Controlled SIDO Buck LED Driver. *IEEE Trans. Power Electron.* **2023**, *38*, 12834–12845. [CrossRef]
10. Zeng, J.; Liu, F.; Liu, J.; Cheng, K.W.E. A Flexible Mode Electrolytic Capacitor-Free LED Driver With High Efficiency Over a Wide Range of Input Voltage. *IEEE Trans. Power Electron.* **2020**, *35*, 8490–8500. [CrossRef]
11. Abdelmessih, G.Z.G.; Alonso, J.M.; Spode, N.d.S.; Costa, M.A.D. High-Efficient electrolytic-capacitor-less off-line led driver with reduced power processing. *IEEE Trans. Power Electron.* **2022**, *37*, 1804–1815. [CrossRef]

12. Cheng, Q.; Liu, J.; Lee, H. A Wide-Input-Range Low-Profile Adaptive Delay Compensated Hysteretic LED Driver with Enhanced Current Accuracy. *IEEE J. Solid-State Circuits* **2022**, *57*, 2474–2485. [CrossRef]
13. Dasohari, M.; Vishwanathan, N.; Porpandiselvi, S.; Vani, A.R.M. A Soft-Switched Boost Converter Based LED Driver with Reduced Input Current Ripple. *IEEE Access* **2024**, *12*, 45904–45922. [CrossRef]
14. Alonso, J.M.; Abdelmessih, G.Z.; Costa, M.A.D.; Guan, Y.; Wang, Y. Investigation of the Use of Switched Capacitor Converters as LED Drivers. *IEEE Trans. Ind. Appl.* **2025**, *61*, 1305–1317. [CrossRef]
15. Qu, Y.; Wang, Z. Soft-Switching Techniques for Single-Inductor Multiple-Output LED Drivers. *IEEE Trans. Power Electron.* **2020**, *35*, 13748–13756. [CrossRef]
16. Ye, C.; Das, P.; Sahoo, S.K. Peak Current Control of Multichannel LED Driver with Selective Dimming. *IEEE Trans. Ind. Electron.* **2019**, *66*, 3446–3457. [CrossRef]
17. Wang, Y.; Gao, S.; Guan, Y.; Huang, J.; Xu, D.; Wang, W. A Single-Stage LED Driver Based on Double LLC Resonant Tanks for Automobile Headlight with Digital Control. *IEEE Trans. Transp. Electrif.* **2016**, *2*, 357–368. [CrossRef]
18. Kolla, H.R.; Vishwanathan, N.; Murthy, B.K. Independently Controllable Dual-Output Half-Bridge Series Resonant Converter for LED Driver Application. *IEEE J. Emerg. Sel. Top. Power Electron.* **2022**, *10*, 2178–2189. [CrossRef]
19. Zhou, S.; Zhou, G.; Liu, G.; Mao, G. Small-Signal Modeling and Cross-Regulation Suppressing for Current-Mode Controlled Single-Inductor Dual-Output DC–DC Converters. *IEEE Trans. Ind. Electron.* **2021**, *68*, 5744–5755. [CrossRef]
20. Fang, P.; Sheng, B.; Webb, S.; Zhang, Y.; Liu, Y.-F. LED Driver Achieves Electrolytic Capacitor-Less and Flicker-Free Operation With an Energy Buffer Unit. *IEEE Trans. Power Electron.* **2019**, *34*, 6777–6793. [CrossRef]
21. He, Q.; Luo, Q.; Huang, J.; Cao, C.; Sun, P. Analysis and design of modular open-loop LED driver with multi-channel output currents. *IET Power Electron.* **2019**, *12*, 1721–1729. [CrossRef]
22. Molavi, N.; Farzanehfard, H. A Nonisolated Wide-Range Resonant Converter for LED Driver Applications. *IEEE Trans. Ind. Electron.* **2023**, *70*, 8939–8946. [CrossRef]
23. Patakamoori, A.; Udumula, R.R.; Nizami, T.K.; Ch, K.R.R. An Efficient Soft-Switched LED Driver for Street Lighting Applications with Input Regulation. *IEEE J. Emerg. Sel. Top. Power Electron.* **2023**, *11*, 5018–5028. [CrossRef]

energies

MDPI

Article

Variable Dimensional Bayesian Method for Identifying Depth Parameters of Substation Grounding Grid Based on Pulsed Eddy Current

Xiaofei Kang [1], Zhiling Li [1], Jie Hou [1], Su Xu [1], Yanjun Zhang [1], Zhihao Zhou [1] and Jingang Wang [2,*]

1 Baotou Power Supply Branch of Inner Mongolia Power (Group) Co., Ltd., Baotou 014030, China; xiaofei_kang1119@163.com (X.K.); zhilingli112@proton.me (Z.L.); jiehou621@proton.me (J.H.); xusu0802@163.com (S.X.); yanjun_zhang_btps@163.com (Y.Z.); zhouzhihao1108@proton.me (Z.Z.)
2 School of Electrical Engineering, Chongqing University, Chongqing 400044, China
* Correspondence: jingang@cqu.edu.cn

Abstract

The substation grounding grid, as the primary path for fault current dissipation, is crucial for ensuring the safe operation of the power system and requires regular inspection. The pulsed eddy current method, known for its non-destructive and efficient features, is widely used in grounding grid detection. However, during the parameter identification process, it is prone to local minima or no solution. To address this issue, this paper first develops a pulsed eddy current forward response model for the substation grounding grid based on the magnetic dipole superposition principle, with accuracy validation. Then, a variable dimensional Bayesian parameter identification method is introduced, utilizing the Reversible-Jump Markov Chain Monte Carlo (RJMCMC) algorithm. By using nonlinear optimization results as the initial model and introducing a dual-factor control strategy to dynamically adjust the sampling step size, the model enhances coverage of high-probability regions, enabling effective estimation of grounding grid parameter uncertainties. Finally, the proposed method is validated by comparing the forward response model with field test results, showing that the error is within 10%, demonstrating both the accuracy and practical applicability of the proposed parameter identification method.

Keywords: substation grounding grid; pulsed eddy current; forward response model; variable dimensional Bayesian; parameter identification

1. Introduction

In recent years, as the scale of power transmission and transformation projects continues to expand to meet the needs of the new power system, the number of substations has been increasing annually. The grounding grid, as a critical component of the substation, is primarily designed to efficiently direct fault current and lightning discharge current into the earth, ensuring the safety of equipment and personnel [1,2]. However, during actual construction, some grounding grids fail to meet design and operational requirements. Typical issues include key parameters, such as the burial depth of grounding bodies, not meeting the relevant standards [3–5], resulting in high grounding resistance that cannot effectively dissipate fault and lightning currents. In severe cases, this may jeopardize the safe and stable operation of the power system. Therefore, there is an urgent need for a fast and reliable method to detect key parameters of the grounding grid to enhance the operational safety and engineering efficiency of substation grounding systems [6,7].

The pulsed eddy current method, known for its excellent vertical resolution, especially its high sensitivity in detecting low-resistance conductors, has been widely used in underground metal target detection in recent years [8,9]. The substation grounding grid, as a typical good conductor, has a resistivity much lower than the surrounding soil medium. The electromagnetic field propagates more quickly through the soil than through the grounding conductor, which allows the pulsed eddy current method to provide high-layer resolution when detecting the grounding grid [10–12]. By obtaining high signal-to-noise ratio pulsed eddy current data, the underground structure of the substation grounding grid can be accurately determined through parameter identification. Parameter identification is the process of inferring the spatial distribution characteristics of underground targets based on observed data, and it is one of the most challenging aspects of pulsed eddy current measurements [13]. Common parameter identification methods include "smoke ring" rapid imaging, least squares method, and machine learning-based algorithms. Among them, the "smoke ring" method is an approximate analytical model based on the assumption of uniform half-space. Its depth inversion depends on the empirical coefficient and is sensitive to the dip angle of the anomalous body, so it is difficult to accurately reveal the depth information of the anomalous body [14,15]. The least squares method constructs a target function to measure the discrepancy between the observed data and the forward model, optimizing it numerically to improve detection efficiency and accuracy [16]. For instance, Wang [17,18] and Chen [19,20] used Differential Evolution (DE) algorithms to enhance the accuracy of parameter identification models. However, the Jacobian matrix will lead to too many conditions due to its ill-conditioned nature. This kind of global optimization algorithm usually has more iterations and poor real-time performance. With the advancement of computational power, machine learning applications in electromagnetic parameter identification have gained momentum. Thomas et al. used machine learning to quickly estimate the burial depth of underground metal materials [21] and further classified metal material types, achieving a recognition accuracy of 90% [22]. Li et al. [23] proposed a parameter identification method based on deep neural networks to reconstruct the electromagnetic response of underground targets, significantly improving the accuracy and computational efficiency of parameter identification. Although machine learning has the advantages of fast speed and high accuracy, it relies on a large number of high-dimensional training data and easily leads to data dimension disaster. At present, there are few practical application cases of the pulsed eddy current method in substation grounding grid detection, and data acquisition is more difficult. The non-training set distribution caused by substation grounding grid corrosion reduces its universality.

This study aims to achieve precise identification of underground substation grounding grids; thus, it proposes high requirements for the parameter identification algorithm, which must possess good global search capabilities and strong noise resistance. To address the non-uniqueness problem in parameter identification, as well as the limitations of traditional methods that only provide a single optimal solution and cannot assess parameter uncertainty, a study on electromagnetic data variable dimensional Bayesian probability parameter identification is conducted. Firstly, in the second section, an electromagnetic forward model suitable for substation grounding grid is constructed as the basis for subsequent parameter identification. Secondly, in the third section, considering the complex transformation characteristics of underground structures in different environments, the Reversible-Jump Markov Chain Monte Carlo (RJMCMC) algorithm is introduced, and the nonlinear optimization parameter identification results are used as the initial model. A two-factor control strategy is proposed to dynamically adjust the sampling step size, so as to enhance the sampling model's ability to cover the high-probability region of the parameter space and realize the effective estimation of the model parameter uncertainty. Finally, the feasibility and effectiveness of the proposed method in substation grounding grid detection are verified by the model and field test in Sections 4 and 5, which shows that the method has good performance in dealing with non-uniqueness and noise interference.

2. Electromagnetic Forward Model of Substation Grounding Grid

2.1. Construction of Electromagnetic Forward Model

When using the pulsed eddy current method to measure the substation grounding grid, it can be considered as an underground low-resistance anomaly. Through steps such as noise reduction and parameter identification, the underground anomaly can be effectively recognized [24]. Due to the different resistivity and thickness between the underground grounding grid and surrounding mediums, the resistivity and thickness of each layer are set as ρ_i and h_i, respectively, as shown in Figure 1. The buried grounding grid is regarded as one layer in the layered model.

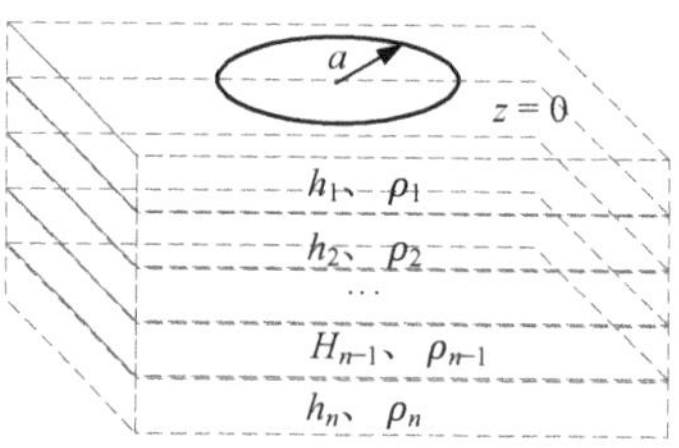

Figure 1. Schematic of the layered model.

In this study, both the transmitting and receiving coils are circular. The coils are designed with circular symmetry, i.e., a central return loop structure, which simplifies the mathematical description and modeling process for forward modeling and parameter identification. The central return loop source device consists of two coils, used as transmitting and receiving magnetic dipoles, respectively. The electromagnetic field components excited by the vertical magnetic dipole source are TE (Transverse Electric)-type fields. The TE, TM (Transverse Magnetic) fields, and the properties of the geological layers differ. The electric field in the TE-type field is parallel to the layer plane, with a strong penetration ability through high-resistance layers and minimal anisotropic influence from the layers. In this paper, a magnetic source is used to detect the grounding system.

At this point, the electromagnetic potential function F is introduced, which conveniently represents the electric field E and the magnetic field H. In the coordinate system of Figure 1, the potential function F of a single magnetic dipole is given by the following expression:

$$F(r,z) = \frac{m}{4\pi}\int_0^\infty \left[e^{-u_i|z+h|} + r_{TE}e^{u_i(z-h)}\right]\frac{\lambda}{u_i}J_0(\lambda r)d\lambda \tag{1}$$

where m represents the magnetic moment of the dipole, $J_0(\lambda r)$ is the zeroth-order Bessel function, indicating the diffusion of the magnetic field in space, λ is the integration variable, describing the propagation characteristics of the electromagnetic field in space. $\lambda = \sqrt{k_x^2 + k_y^2}$, k_x and k_y represent the wave numbers in the x- and y-directions, respectively. $u_n = \sqrt{\lambda^2 - k_n^2}$, ${k_n}^2$ represents the wave numbers of the n-th layer, and u_0 represents the permeability of air. The variable r is the radial distance from the source to the receiver, z is the vertical coordinate for the layered medium, and r_{TE} is the reflection coefficient. The coil is radially divided into numerous tiny current loops, and each loop is regarded as a magnetic dipole (the magnetic moment mis proportional to the current I and the area). By integrating (1) within the range of radius a and using the recurrence relation of Bessel functions, (2) containing the first-order Bessel function $J_1(\lambda r)$ is finally obtained:

$$F(r,z) = \frac{Ia}{2}\int_0^\infty \frac{1}{u_i}\left[e^{-u_i|z+h|} + r_{TE}e^{u_i(z-h)}\right]J_1(\lambda a)J_0(\lambda r)d\lambda \tag{2}$$

where $J_1(\lambda r)$ is the first-order Bessel function used to describe the current distribution in the return loop source. In cylindrical coordinates, when located on the ground surface (i.e., $z = 0$, $h = 0$), the magnetic field and electric field generated at the center of the horizontal central return loop transmitting coil are expressed as follows:

$$\begin{cases} E_\varphi = -i\omega\mu\frac{\partial F}{\partial r} = -\frac{\widehat{z}_0 Ia}{2}\int_0^\infty[1+r_{TE}]\frac{\lambda}{u_0}J_1(\lambda a)J_1(\lambda r)d\lambda \\ H_r = \frac{\partial^2 F}{\partial r\partial z} = \frac{Ia}{2}\int_0^\infty[1-r_{TE}]\lambda J_1(\lambda a)J_1(\lambda r)d\lambda \\ H_z = -\frac{1}{r}\frac{\partial F}{\partial r}(r\frac{\partial F}{\partial r}) = \frac{Ia}{2}\int_0^\infty[1+r_{TE}]\frac{\lambda^2}{u_0}J_1(\lambda a)J_0(\lambda r)d\lambda \end{cases} \tag{3}$$

where $\widehat{z}_0 = i\omega\mu_0$. At the ground surface ($z = 0$), in the expressions of the magnetic field Hz and the electric field E at the center of the horizontal loop coil, the reflection coefficient r_{TE} directly reflects the difference in electromagnetic characteristics between the underground media (soil and grounding grid); the grounding grid, as a low-resistivity body, will increase the modulus of r_{TE}, resulting in an increase in the rate of change in the magnetic field amplitude in Equation (3). The reflection coefficient r_{TE} is calculated as

$$r_{\mathrm{TE}} = \frac{\lambda - \hat{u}_1}{\lambda + \hat{u}_1} \tag{4}$$

When the magnetic source serves as the excitation source, Equation (4) can be transformed into

$$r_{TE} = \frac{Y_0 - \hat{Y}_1}{Y_0 + \hat{Y}_1} \tag{5}$$

In Equation (5), Y_0 and $\widehat{Y}_1$ describe the propagation impedance of the electromagnetic field in the layered medium, which can be obtained by the recursive relationship in Equation (6):

$$\begin{cases} Y_0 = \frac{u_0}{\widehat{z}_0} \\ \widehat{Y}_1 = Y_1\frac{\widehat{Y}_2 + Y_1\tanh(u_1h_1)}{\widehat{Y}_1 + Y_2\tanh(u_1h_1)} \\ \widehat{Y}_n = Y_n\frac{\widehat{Y}_{n+1} + Y_n\tanh(u_nh_n)}{\widehat{Y}_n + Y_{n+1}\tanh(u_nh_n)} \\ \widehat{Y}_n = Y_N = \frac{u_n}{i\omega u_0} \end{cases} \tag{6}$$

To simplify the calculations, only the center point of the received component is considered, where $r = 0$, while $J_0(0)$ and $J_1(0)$ are 1 and 0, respectively. The electromagnetic field expression at the center location is

$$\begin{cases} E_\phi = 0 \\ H_r = 0 \\ H_z = \frac{Ia}{2}\int_0^\infty[1+r_{TE}]\frac{\lambda^2}{u_0}J_1(\lambda a)d\lambda \\ \frac{dB_z(\omega)}{dt} = -i\omega\mu_0\frac{Ia}{2}\int_0^\infty[1+r_{TE}]\frac{\lambda^2}{u_0}J_1(\lambda a)d\lambda \end{cases} \tag{7}$$

The magnetic field at the center point, denoted as Hz, is the frequency-domain response $H_z(\omega)$. Since the measurement of the pulsed eddy current method primarily involves parameter identification based on induced electromotive force, the derivative of $H_z(\omega)$ with respect to time gives the fourth term in Equation (7). To obtain the time-domain electromagnetic field response, a Fourier transform of Equation (7) is performed. During the derivation process, the Fourier transform is converted into the integral of high-oscillation sine and cosine functions as follows:

$$\begin{aligned} f_c(k) &= \int_0^{+\infty} F(x)\cos(kx)dx \\ f_s(k) &= \int_0^{+\infty} F(x)\sin(kx)dx \end{aligned} \tag{8}$$

In Equation (8), due to the high oscillation of sin(kx) and cos(kx), the numerical solution obtained by approximate integration often does not meet the required precision. Researchers have found that for most practical problems, including the substation grounding grid in this study, electromagnetic response cannot be directly solved by an analytical solution, and numerical integration methods are required in the forward calculation process. For example, due to the oscillatory nature of Bessel functions, especially at relatively high wave numbers, numerically performing the integration is extremely complex. A commonly used approach nowadays is to carry out approximate calculation through the fast Hankel transform.

After obtaining the frequency-domain solution of the vertical magnetic field, the usual practice is to obtain the time-domain solution by converting the digital filtering operator. This paper first uses the filter coefficients from the Hankel transform of the first-order Bessel function to calculate the electromagnetic signal in the frequency domain [25]. The calculation expression is as follows:

$$\begin{cases} f(r) = \dfrac{1}{r}\sum\limits_{i=1}^{n} K(\lambda_i)W_i \\ \lambda_i = \dfrac{1}{r}10^{[a+(i-1)s]}, i = 1,2,\cdots,n \end{cases} \tag{9}$$

where i represents the sampling point location and W_i is the filter coefficient. From Equation (7), it is known that linear computation of the first-order Bessel function is required, with $n = 140$, $a = -7.91001919000$, and $s = 8.7967143957 \times 10^{-2}$. Using the above expression, the electromagnetic signal in the frequency domain can be calculated.

After calculating the frequency-domain signal, since the subsequent signal processing in this paper involves time-domain signal processing, it is necessary to convert the frequency-domain signal into a time-domain signal. By using the sine and cosine transform method, the frequency-domain signal $H_z(\omega)$ is transformed into a time-domain signal, and the expressions for the vertical magnetic field time-domain signal and the derivative of the magnetic field in terms of the sine–cosine transform are as follows:

$$\begin{cases} H_z(t) = -\dfrac{2}{\pi}\int_0^{\infty} Im[H_z(\omega)]\dfrac{1}{\omega}\cos(\omega t)d\omega \\ = H_z(0) - \dfrac{2}{\pi}\int_0^{\infty} Re[H_z(\omega)]\dfrac{1}{\omega}\sin(\omega t)d\omega \\ \dfrac{\partial H_z(t)}{\partial t} = \dfrac{2}{\pi}\int_0^{\infty} Im[H_z(\omega)]\sin(\omega t)d\omega \\ = -\dfrac{2}{\pi}\int_0^{\infty} Re[H_z(\omega)]\cos(\omega t)d\omega \end{cases} \tag{10}$$

The time-domain signal can be obtained by the sine transform method as follows:

$$f_s(t) = \frac{1}{t}\sqrt{\frac{\pi}{2}}\sum_{n=N_1}^{N_2} F(\omega)c\sin(n\Delta) \tag{11}$$

where Δ is the sampling interval and $c\sin(n\Delta)$ is the transform filter coefficient, discretizing $\omega = e^{n\Delta}/t$ and substituting it into the electromagnetic field response in the frequency domain for calculation. Finally, the response $V(t)$ at the center of the layered earth model for the central return loop device is

$$V_1(t) = q\mu_0 \frac{\partial H_z(t)}{\partial t} \tag{12}$$

where μ_0 represents the permeability of free space, and q is the equivalent area of the excitation coil. After the central return loop generates an electromagnetic signal through the magnetic source, the magnetic field response from the grounding grid is collected by the receiving coil. The induced voltage $V_1(t)$ of the receiving coil is proportional to the time derivative of the magnetic field ($\partial Hz/\partial t$), and the burial depth of the grounding grid will affect the diffusion speed of the magnetic field: the greater the burial depth, the longer the time delay for the magnetic field to reach the ground surface. An increase in the number of turns n of the coil and the effective area S in (12) will increase the amplitude of $V_1(t)$. Essentially, this corresponds to obtaining the transfer function of the earth. The collected signal is expressed as the induced voltage form $V_2(t)$:

$$V_2(t) = n\frac{d\varphi}{dt} = nS\frac{dB}{dt} \tag{13}$$

In Equation (13), n and S are the number of turns and effective area of the receiving coil, respectively, and B is the magnetic flux density along the axis of the receiving coil. In the parameter identification algorithm, the relationship between the induced voltage signal collected by the receiving coil and the induced voltage signal calculated by the forward algorithm is used to determine the structure of the underground medium.

2.2. Verification of Forward Model

An H-type model for the substation grounding grid is constructed, where the resistivity of the middle layer, representing the substation grounding grid, is significantly lower than the resistivity of the upper and lower soil layers, as shown in Figure 2. The model parameters are set as listed in Table 1. The excitation current amplitude is set to 20 A, with the number of turns for the transmitting coil set to 10 and a radius of 0.4 m. An H-type model for the substation grounding grid is constructed, and its specific structure is shown in Figure 2. The model divides the underground structure of the grounding grid into three layers, which are soil, grounding grid, and soil from bottom to top. The settings of various parameters of the model are detailed in Table 1, where Resistivity represents the resistivity of each layer, Thickness represents the thickness of each layer, and ρ_2 and h_2 correspond to the resistivity and thickness of the grounding grid, respectively. The basis for determining the parameter values is explained as follows: The main material of the grounding grid is galvanized steel, whose conductivity is usually 1×10^6 S/m or higher. Considering that this conductivity value is significantly different from that of soil, and when the conductivity is greater than 1×10^6 S/m, it has little impact on the calculation results, 5.6×10^6 S/m is selected in this paper for conversion to determine ρ_2. In addition, the thickness h of the grounding grid is set to 0.012 m, which is in line with the conventional value range for grounding grid design in China. The excitation current amplitude is set to 20 A, with the number of turns for the transmitting coil set to 10 and a radius of 0.4 m.

The number of turns for the receiving coil is set to 100, with a radius of 0.1 m. The resulting forward modeling results are shown below.

Figure 3 shows the variation in the pulsed eddy current signal during the grounding grid detection process, plotted on a logarithmic scale. The inset uses a linear scale for the x-axis and a logarithmic scale for the y-axis to facilitate observation of early response details. In the H-type geo-electric model, due to the resistivity of the layers exhibiting a high–low–high variation trend, the pulsed eddy current response curve shows a fast–slow–fast attenuation pattern. However, because the grounding grid being detected in this study is very small, the attenuation variation trend is not very obvious, and the dynamic range is extremely large. This further increases the difficulty of parameter identification. In common parameter identification algorithms, it is even possible for the grounding grid to be overlooked. Therefore, in the parameter identification algorithm used in this paper, prior information from the grounding grid design is incorporated through additional conditions to assist the parameter identification process.

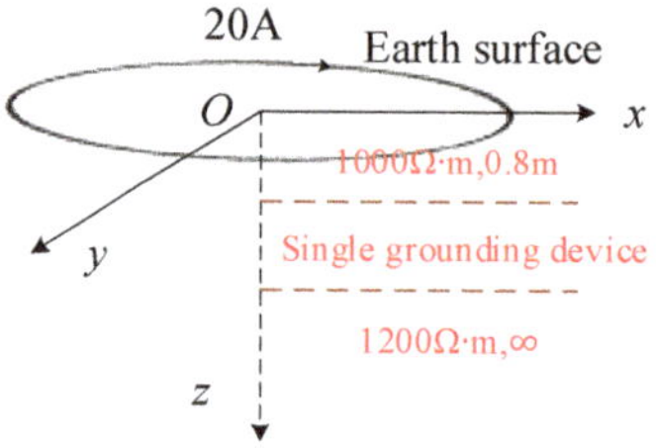

Figure 2. H-type grounding grid electromagnetic model.

Table 1. H-type model parameters.

Resistivity	Unit (Ω·m)	Thickness	Unit (m)
ρ_1	900	h_1	0.8
ρ_2	$1/(5.6 \times 10^6)$	h_2	0.012
ρ_3	1000	h_3	∞

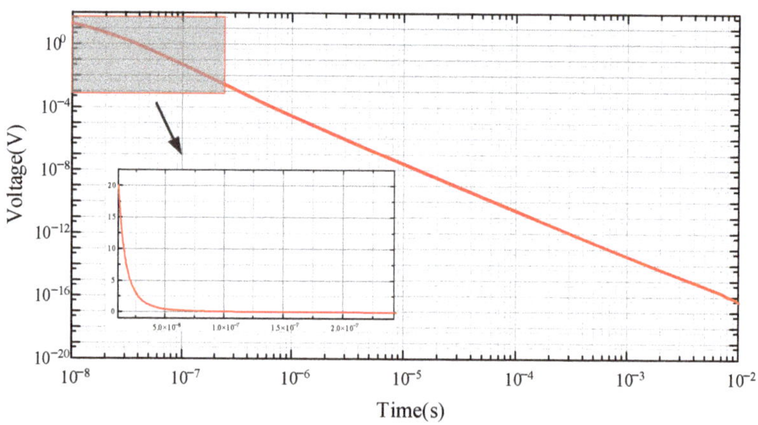

Figure 3. Pulse eddy current forward signal of H-type grounding grid.

3. Substation Grounding Grid Parameter Identification Method

The goal of pulsed eddy current parameter identification is to estimate the model parameters m that best fit the real conditions of the grounding grid based on the observed data d. However, due to the uncertainty of the observed data, parameter identification is often challenged by non-uniqueness and data errors caused by noise. Even with the same observed data, multiple reasonable underground structural models may correspond, making it difficult to fully describe model uncertainty by relying solely on traditional least squares fitting or regularization methods. To more reasonably characterize the uncertainty in the parameter identification results and fully utilize available geological information, Bayesian parameter identification is introduced, as shown in the following equation:

$$P(m|d) = \frac{P(d|m)P(m)}{P(d)} \tag{14}$$

Under the Bayesian parameter identification framework, the solution for the geo-electric model parameters m follows Bayes' theorem, where the posterior probability distribution $P(m \mid d)$ is estimated given the observed data d. Here, $P(m \mid d)$ represents the posterior probability distribution, which evaluates the probability of each possible geo-electric model m after measuring the electromagnetic data d. This is the primary goal of Bayesian inference. $P(d \mid m)$ is the likelihood function, which describes the probability of observing the electromagnetic data d given the geoelectric model m, indicating how well the model fits the observed data. $P(m)$ is the prior distribution, reflecting the prior expectations of the model parameters before incorporating the observed data. In this paper, the prior information is constrained based on national standards and field measurement data. $P(d)$ is the marginal likelihood, which serves as the normalization factor to ensure the posterior probability density is normalized. Its value is calculated by integrating over all possible geo-electric models' contributions to the data d, as shown below:

$$P(d) = \int P(d|m)P(m)dm \tag{15}$$

The Bayesian parameter identification method optimizes and constrains the model parameters by combining prior information with the degree of fit to the observed data, operating within a probabilistic statistical framework. In the identification of depth parameters of substation grounding grids, the strong non-uniqueness of geophysical parameter identification makes it difficult for a single optimal solution to fully reflect model uncertainty. However, the Bayesian method can quantify uncertainty by providing the probability distribution of model parameters, thereby improving the robustness and physical interpretability of the results. Nevertheless, in the practical application of electromagnetic identification, the Bayesian method has two main problems.

First, the calculation of the posterior probability distribution is difficult. The posterior probability is derived from the product of the prior distribution and the likelihood function, and needs to be normalized over the entire model space, which involves high-dimensional integration and is usually difficult to solve analytically. The electromagnetic parameter identification process in this paper also has nonlinear and high-dimensional characteristics, so a sampling method instead of integration is adopted, and a large number of statistical samples are used to approximate the optimal solution. Second, sample generation is difficult. To approximate the posterior probability distribution, a large number of samples need to be generated from the model parameter space. The common Markov Chain Monte Carlo (MCMC) method can extensively and randomly search for model parameters based on prior information and generate predicted data close to the observed data. However, when directly applied to high-dimensional spaces, its efficiency is extremely low. Moreover, the traditional MCMC method has a fixed parameter dimension, which makes it difficult to adapt to changes in the number of layers and interface positions in the grounding body model. To address these issues, the Reversible-Jump Markov Chain Monte Carlo (RJMCMC) method has been introduced. This method enables dynamic adjustment of the dimensions of model parameters (such as the number of layers and interface positions) through jump sampling between different dimensions, thereby exploring the possible model space more efficiently and obtaining a posterior probability distribution more consistent with the observed data. It effectively improves sampling efficiency and enhances the accuracy of complex Bayesian parameter identification.

3.1. Model Parameter Initialization

In the electromagnetic model of the substation grounding grid, the key model parameters mainly include resistivity, number of layers, and interface depth. The initial model can either be specified by the user or randomly generated based on certain rules. A reasonable initial model setup helps to start sampling from a parameter space that is closer to the true solution, thus effectively reducing the number of ineffective samples during the early stages of the MCMC algorithm.

Since MCMC sampling often starts from models with low posterior probability, it requires numerous iterations to gradually approach high-posterior regions. This phase is known as the burn-in phase. The length of the burn-in phase is influenced by multiple factors, including the complexity of the substation grounding grid parameter identification problem, the quality of the initial model, sampling strategy, and convergence criteria. Selecting a more reasonable initial model can significantly shorten the burn-in process and improve overall parameter identification efficiency.

In this study, prior to parameter identification, an optimized strategy is introduced to construct an approximately high-quality initial model to accelerate the convergence process during the burn-in phase. It should be emphasized that this initial model is only used to optimize sampling efficiency and is not intended to serve as the final parameter identification solution. Inspired by the "snow ablation" mechanism in the Snow Ablation Optimizer (SAO), which exhibits a good balance between global search and local exploitation (Exploration and Exploitation, ENE), this approach can generate an initial solution that is closer to the true model distribution, thereby improving the overall efficiency of the RJMCMC sampling process.

3.2. Prior Function and Likelihood Function Generation

In Bayesian parameter identification, the prior function and likelihood function are central to the calculation of the posterior probability distribution. They collectively determine the estimated results of the model parameters.

(1) Prior Function

During the model parameter setting process, parameter thresholds need to be set based on prior information. For the resistivity search range, it is typically set based on known geological information, as different substation grounding grids have varying resistivity ranges for soil environments. Constraints are applied to resistivity, the number of layers, interface depth, and noise-influenced parameters:

(1) Minimum and maximum model layers ($k_{\min}, k_{\max}$).

(2) Minimum and maximum interface depth ($z_{\min}, z_{\max}$).

(3) Minimum and maximum resistivity ($\rho_{\min}, \rho_{\max}$).

(4) Noise impact range $dn(dn_{\min}, dn_{\max})$.

During model perturbation, except for the noise impact, the remaining prior information uses logarithmic parameters to ensure that all parameters are positive and that their perturbation range is more flexible. Given that the characteristics of these four parameters differ, the prior probability calculation should be performed by distributing the calculation, expressed as

$$p(m) = p(k) \cdot p(z|k) \cdot p(\rho|k) \cdot p(dn) \tag{16}$$

Compared to conventional underground model measurements, the object of detection in this paper is the substation grounding grid. Therefore, a fixed parameter layer can be set, where the thickness and resistivity of the grounding grid are predetermined. In this case, $p(k)$ is not affected by the layer of the underground grounding grid and is still considered a bounded uniform distribution:

$$p(k) = \frac{1}{k_{\max} - k_{\min} + 1}, k_{\min} \leq k \leq k_{\max} \tag{17}$$

Before performing signal parameter identification, the surface resistivity can first be estimated via parameter identification. The surface resistivity ρ_0 measured at this point can be used as the initial mean to construct a Gaussian distribution. A weight parameter is introduced to adjust the influence of the prior distribution on the final parameter identification results. The expression for $p(\rho \mid k)$ is as follows:

$$p(\rho|k) = \frac{1}{\left[(2\pi)^k |\mathbf{C}_\rho|\right]^{\frac{1}{2}}} \exp\left[-\frac{\omega}{2}(\rho - \rho_0)^{\mathrm{T}} \mathbf{C}_\rho^{-1} (\rho - \rho_0)\right] \tag{18}$$

where ω is the weight parameter. The larger ω is, the stronger the influence of the prior distribution, and the parameter identification results will tend to align with the prior information. Conversely, smaller ω values rely more on the observed data. $C\rho$ is the variance matrix of the resistivity prior information, which is a diagonal matrix, with $C_\rho = \mathrm{diag}(\log(1 + \rho_r)^2)$. Due to the existence of the grounding grid, the magnitude difference between resistivities is vast, so logarithmic operations are used in the calculation. When the number of layers k is determined, there are $k!$ ways to arrange each layer's interface at selected depths. Because of the grounding grid, its thickness and distribution can be considered a fixed value, and thus, $p(z \mid k)$ in this section can be modified as

$$p(z|k) = \frac{1}{\Delta z^{k-1}}, z_{\min} \leq z_i \leq z_{\max}, \forall i = 1, 2, \cdots, k \tag{19}$$

where Δz is the layer thickness, and noise signals are assumed to follow a uniform distribution:

$$p(dn) = \frac{1}{dn_{\max} - dn_{\min}}, dn_{\min} \leq dn \leq dn_{\max} \tag{20}$$

Given the presence of the underground grounding grid, the number of initial model layer interfaces for parameter identification must be at least 2, i.e., there must be 3 layers underground.

(2) Likelihood Function

In Bayesian parameter identification, the likelihood function primarily serves to assess the degree of fit between the candidate model m that generates the data d_{model} and the actual observed data d. The higher the likelihood value, the better the model fits the observed data. After electromagnetic data denoising, errors due to measurements and prior signal processing can be considered Gaussian, and the likelihood function can be expressed as:

$$\begin{aligned} p(\boldsymbol{d}|k, z, \boldsymbol{\rho}, dn) &= \frac{1}{\sqrt{(2\pi)^{N_d}|C_d|}} \\ \exp&\left[-\frac{\omega}{2}(\boldsymbol{d} - f(\boldsymbol{m}))^{\mathrm{T}} C_d^{-1} (\boldsymbol{d} - f(\boldsymbol{m}))\right] \end{aligned} \tag{21}$$

where d represents the observed data, $f(m)$ is the forward operator, and C_d is the prior variance matrix. The diagonal elements of C_d represent the measurement uncertainty, i.e., variance, and it is assumed that the data errors are independent. The larger C_d is, the greater the measurement error. In such cases, the impact of prior information should be increased, and thus, a weight factor ω is added to control the influence of prior information on sample generation. N_d represents the number of observed data points.

3.3. RJMCMC Method Improvement

When the number of layers in the subsurface is determined or the underground structure is simple, the samples generated by MCMC at the theoretical level can identify a set of optimal grounding grid models to simulate the underground structure. However, due to the large magnitude difference between the resistivity of the grounding grid and other underground media, simply dividing the underground structure into three layers can easily lead to multiple high-quality solutions, which do not meet the required results for parameter identification, as described in Section 2. Furthermore, if the outdoor measurement environment is complex, such as in mountainous areas, it significantly reduces the measurement accuracy. Therefore, during the sampling process, the number of samples produced in the initial stage should be as rich as possible, and during the iterations, the number of layers and the location of interfaces in the model should be adjusted based on the results. The RJMCMC sampling process is similar to MCMC, but RJMCMC allows sampling to jump between different model dimensions to better explore the underground model of the grounding grid.

The main purpose of RJMCMC sampling is to find an optimal underground electrical model that best fits the observed data. However, due to the non-uniqueness of geophysical parameter identification, there are usually multiple possible models that can explain the same data. Therefore, the goal of this paper is not to find a unique solution but to sample from all possible models and ultimately select the model set that best fits the observed data. RJMCMC introduces a dimensionality-changing mechanism, allowing the model to freely jump between different model spaces, thus flexibly searching for the optimal solution. Additionally, to improve sampling accuracy and efficiency, a dual-factor (ω-c) is introduced to enhance the resolution of resistivity and dynamically control the sampling step size.

(1) Model Generation Based on Dual-Factor Control

Based on the horizontal layered model of the grounding grid, an iterative method using dual-factor control (ω-c) is proposed, which controls the weight of prior information and the step size of sampling to construct a complete electromagnetic data distribution model. In the proposal distribution sampling stage, candidate models are built for the three main parameters of parameter identification (resistivity, interface depth, and number of layers). Candidate models are typically generated in the following two ways: RJMCMC uses the Birth/Death mechanism to adjust the number of layers and the depth of interfaces; after determining the number of layers and interfaces, the resistivity for each layer is adjusted. Therefore, the proposal distribution function $q(m' \mid m)$ in the RJMCMC sampling process consists of the following three parts:

$$q(m'|m) = q(k'|k) \cdot q(z'|k',z) \cdot q(\rho'|k',z',\rho) \tag{22}$$

where $q(k' \mid k)$ controls the change in the number of layers through the Birth/Death mechanism, $q(z' \mid k', z)$ controls the adjustment of the interface depth after the number of layers is determined, and $q(\rho' \mid k', z', \rho)$ controls the adjustment of resistivity. The update of the resistivity parameter is determined by the proposal distribution and follows a Gaussian distribution. That is, the new resistivity value is randomly drawn from a Gaussian distribution with the current model's resistivity as the mean. Therefore, except for the grounding grid layer, the expression is similar to that in the prior Equation (18).

$$q(\rho'|k',z',\rho) = \frac{1}{\left[(2\pi)^k \left|\mathbf{C}_{\rho_k^{k',z'}}\right|\right]^{\frac{1}{2}}} \exp\left[-\frac{1}{2}\left(\rho_{k'} - \rho_k^{k',z'}\right)^{\mathrm{T}} \mathbf{C}^{-1}_{\rho_k^{k',z'}} \left(\rho_{k'} - \rho_k^{k',z'}\right)\right] \quad (23)$$

where $\rho_{k'}$ represents the resistivity of the candidate model, and the mean vector $\rho_k^{k',z'}$ is the resistivity of the current model, both of which are calculated logarithmically. $\mathbf{C}_{\rho_k^{k',z'}}$ is the posterior covariance matrix, which is typically linearized from the posterior model variance:

$$\mathbf{C}_{\rho_k^{k',z'}} \approx \left[\omega \mathbf{C}_\rho^{-1} + \mathbf{J}^T \mathbf{C}_d^{-1} \mathbf{J}\right]^{-1} \quad (24)$$

Here, **J** is the sensitivity matrix, representing the impact of model parameter changes on the observed data, thus adjusting the sampling step size for model parameters. If a particular value in the matrix is large, it indicates that changes in that resistivity significantly affect the observed data, making the parameter identification result sensitive to this parameter. The physical meaning of ω is consistent with the likelihood function (21), used to balance the weight between model constraints and data fitting. The sampling step size is controlled by the posterior covariance matrix, with a scaling factor c dynamically adjusting the step size:

$$\mathbf{C}'_{\rho_k^{k',z'}} = c \times \mathbf{C}_{\rho_k^{k',z'}} \quad (25)$$

where c is a dynamic factor constructed by sine and cosine functions, which is used to control the scaling of the sampling step size, which equals c_1 plus c_2, as shown in (26).

$$\begin{cases} c_1 = \sin\left(\frac{3\pi * it}{T}\right) + 1.5 \\ c_2 = \cos\left(\frac{3\pi * it}{T}\right) + 1.5 \end{cases} \quad (26)$$

At this point, RJMCMC updates the model in four ways:

(1) Layer Addition (Birth Move): Here, $k' = k + 1$, and a random position is chosen between the minimum depth $z_{\min}$ and maximum depth $z_{\max}$ as the new interface. This interface follows a uniform distribution $U(z_{\min}, z_{\max})$ and forms a new layer. The resistivity of this layer follows a normal distribution $N(\rho, \sigma_{\mathrm{birth}}^2)$, where ρ is the resistivity of the current interface.

(2) Layer Deletion (Death Move): Here, $k' = k - 1$, and a layer is randomly removed from the geo-electric model.

(3) Layer Perturbation (Perturbation Move): A random layer interface is selected for a position change, with the change step size following a normal distribution $N(0, \sigma_{\mathrm{move}}^2)$.

(4) Resistivity Update (Resistivity Change): Here, $k' = k$, $z' = z$, and only the resistivity is updated, with the change step size following a normal distribution $N(\rho, \sigma_{\mathrm{change}}^2)$, where ρ is the current resistivity of the interface.

It should be noted that since the thickness and resistivity of the grounding grid are much smaller than those of the generated model, during model update, it is first checked whether the layer thickness is greater than or equal to 0.1, and whether the resistivity is greater than 0 (logarithmic form is used in the calculation). In the model changes above, the update mode is chosen by randomly generating a number between 0 and 1 to satisfy the following condition:

$$q(k'|k) = \begin{cases} \frac{1}{6}, & k' = k+1\ (\textbf{\textit{Birth}}) \\ \frac{1}{6}, & k' = k-1\ (\textbf{\textit{Death}}) \\ \frac{1}{6}, & k' = k\ (\textbf{\textit{Perturbation}}) \\ \frac{1}{2}, & k' = k\ (\textbf{\textit{Change}}) \end{cases} \tag{27}$$

Because the changes in the number of layers and interface depth significantly impact the model, excessive frequency of layer adjustments during sampling may prevent RJMCMC sampling from converging. Excessive interface perturbations could lead to model instability. Therefore, the sampling probabilities for the first three disturbance types (Birth, Death, Perturbation) are set to $1/6$, preventing overly frequent adjustments in the number of layers. The probability for resistivity updates is increased to $1/2$ to optimize the resistivity distribution while keeping the number of layers fixed. The schematic of state changes is shown below:

Figure 4 demonstrates the four possible states of the new model generated by RJMCMC sampling: (a) layer addition, (b) layer deletion, (c) layer perturbation, and (d) resistivity update. However, not all generated models are suitable for sampling data. Therefore, during the sampling process, each generated model is evaluated for its applicability based on a probability model. This process ensures that the model reasonably fits the observed data and improves the stability and reliability of the parameter identification results.

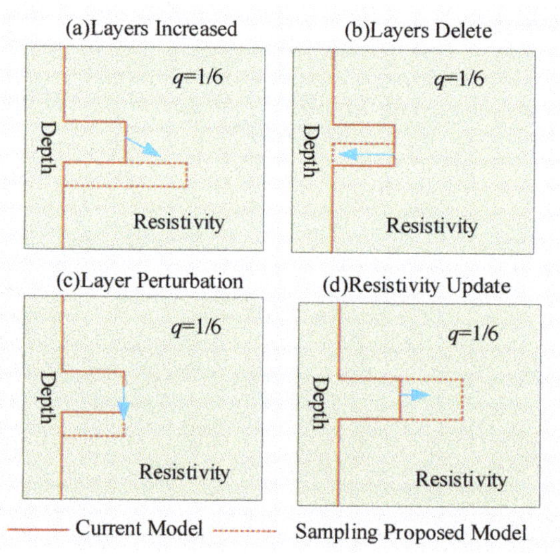

Figure 4. Schematic of four possible states of the new model generated by RJMCMC sampling.

(2) Model Acceptance Probability Setup

Whenever a model is generated, an acceptance–rejection parameter needs to be set to determine if the model should be accepted. According to the Metropolis–Hastings sampling concept, the acceptance probability $\alpha(m' \mid m)$ is calculated as follows:

$$\alpha(m'|m) = \min\left[1, \frac{p(d|m')}{p(d|m)} \cdot \frac{p(m')}{p(m)} \cdot \frac{q(m|m')}{q(m'|m)} \cdot |J|\right] \tag{28}$$

where J is the Jacobian matrix; in this paper, $|J|$ is set to 1 to ensure that the transformation between different dimensions meets the invertibility requirement. The terms $p(d \mid m')/p(d \mid m)$, $p(m')/p(m)$, and $q(m \mid m')/q(m' \mid m)$ are the likelihood ratio, prior probability ratio, and proposal probability ratio, respectively. The prior probability ratio and likelihood ratio are calculated using Equations (16) and (21). The proposal probability ratio measures the probability of transitioning from the current model m to the candidate model m' and is calculated as

$$\frac{q(m|m')}{q(m'|m)} = \begin{cases} \frac{\Delta z\sqrt{2\pi}\sigma_{birth}}{k+1} \exp\left[\frac{(\rho'-\rho)^2}{2\sigma_{birth}^2}\right], \textit{Birth} \\ \frac{k}{\Delta z\sqrt{2\pi}\sigma_{death}} \exp\left[\frac{(\rho'-\rho)^2}{2\sigma_{death}^2}\right], \textit{Death} \\ 1, \textit{Perturbation} \\ 1, \textit{Change} \end{cases} \tag{29}$$

A random number ξ between 0 and 1 is generated. If ξ is smaller than the value of Equation (28), the model is accepted; otherwise, it is rejected, and a new candidate model is generated for sampling, continuing until convergence or the maximum sampling threshold is reached.

3.4. Convergence Criterion for Uncertainty Assessment

To avoid the use of excessive computational resources, we define the sampling convergence condition as being reached when the model's credible interval and the standard deviation of the error's standard deviation remain sufficiently stable during the sampling iterations. The stability condition for the model's credible interval is set as follows:

$$\left|\frac{CR(S_i) - CR(S_i - S_0)}{CR(S_i)}\right| < k_1\% \tag{30}$$

where $CR(x)$ is the boundary value of the credible interval for the x-th sampling, S_i is the current cumulative number of samples, excluding the burn-in phase samples, and S_0 is the pre-set interval of sampling. The physical meaning behind Equation (30) is that when the fluctuation of the model's credible interval range for the S_i-th sampling, compared to the credible interval range of the previous S_0 samples, is less than k_1%, it can be considered that the model's credible interval stability condition has been achieved.

The stability condition for the standard deviation of the error's standard deviation is, specifically, that the system computes the statistical characteristics of the model fitting error every *N_refresh* iterations and maintains a sliding window *N_end* during the sampling process to record the root mean square (RMS) error for the most recent N iterations. Let the error samples in the sliding window be $\{RMS_1, RMS_2, \ldots, RMS_N\}$; then, the corresponding error standard deviation is

$$\sigma_{RMS} = \sqrt{\frac{1}{N-1}\sum_{i=1}^{N}(RMS_i - \overline{RMS})^2} \tag{31}$$

The system further monitors the fluctuation of the error standard deviation across multiple cycles. Let σ_{RMS} calculated over the past several cycles be $\{\sigma_1, \sigma_2, \ldots, \sigma_N\}$. The standard deviation of these standard deviations is defined as

$$\sigma_{std} = \sqrt{\frac{1}{M-1}\sum_{j=1}^{M}(\sigma_j - \overline{\sigma})^2} \tag{32}$$

where $M = N\_end/N\_refresh$. When σ_{std} is smaller than a pre-set threshold δ, it is considered that the sampling process has entered the stable phase, signaling the "end of sampling." In addition, to prevent sampling from getting stuck in low-posterior regions and causing prolonged ineffective sampling, the system sets an upper limit for the number of consecutive rejected models. Once this limit is reached, the sampling process is automatically interrupted to avoid wasting computational resources.

Based on the above analysis, the dimensional Bayesian flowchart is shown in Figure 5.

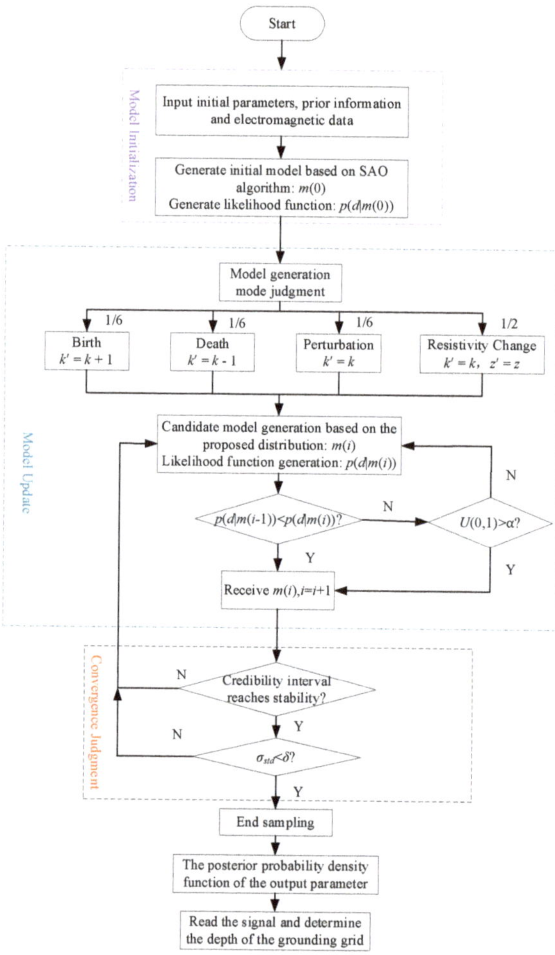

Figure 5. Flowchart for parameter identification of substation grounding grid electromagnetic data.

4. Simulation Model Verification of Parameter Identification

4.1. Parameter Identification Results for the H-Type Grounding Grid Model

The central return loop device is used to perform forward analysis on the H-type grounding body model, as shown in Figure 2. Gaussian random noise with a 5% amplitude ratio is added to the forward results to simulate real-world data. Dimensional Bayesian parameter identification is then performed, with parameter settings as shown in Table 2. All model training and testing were conducted on a workstation equipped with an NVIDIA RTX 3090 GPU (24 GB VRAM), an Intel i9-13900K CPU, and 64 GB RAM.

Table 2. Parameter settings for dimensional Bayesian parameter identification.

Parameters	Values
Minimum Number of Model Layer Interfaces	3
Maximum Number of Model Layer Interfaces	10
Number of Burn-in Phases	10,000
Total Number of Samples	1,000,000
Number of Markov Chains	1
Resistivity Update Step Size σ_{change}	4
Layer Interface Perturbation Step Size σ_{move}	10
New Layer Generation Step Size σ_{birth}	4

The resulting parameter identification results are shown in Figure 6.

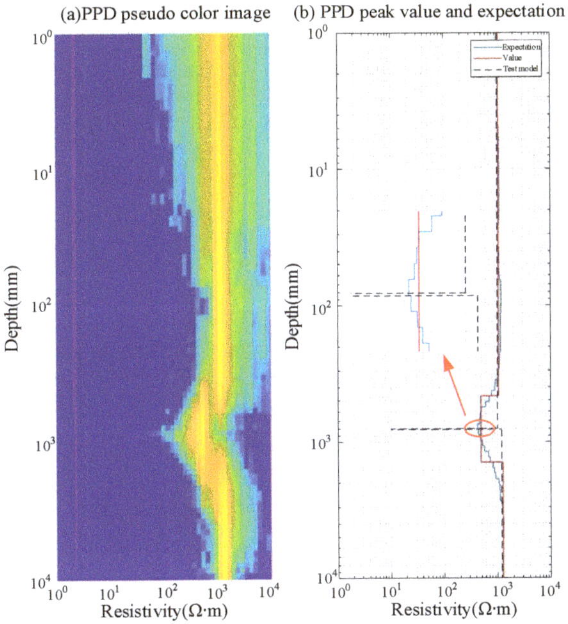

Figure 6. Dimensional Bayesian parameter identification results for H-type grounding grid model.

In Figure 6, PDD represents the posterior probability density $P(m \mid d)$. From (b), it can be seen that the grounding grid's position range is [739.072, 811.131] mm, while the forward model's range is [800, 812] mm. In actual working conditions, the depth of the grounding grid is calculated based on the side of the grounding grid closest to the ground. However, since the Bayesian parameter identification results provide a general range for parameter identification, the measurement error is calculated using the boundary values of both intervals. Specifically, the parameter identification range is set as [h_1, h_2], and the forward model's range average is [h_3, h_4]. The MSE_h (Mean Squared Error) calculation formula is given by

$$MSE\_h = \left(\left|\frac{h\_1 - h\_3}{h\_3}\right| + \left|\frac{h\_2 - h\_4}{h\_4}\right|\right)/2 \times 100\% \tag{33}$$

Using Equation (33), the error is found to be 3.86%, with a central distance of approximately 3.1 cm. The obtained result meets the measurement requirements.

4.2. Parameter Identification Results for Grounding Grid Depth Variation

As shown in Figure 7, electromagnetic parameter identification for the variation in grounding grid depth is conducted, with the input model parameters shown in Table 2. The obtained depth information for the grounding grid is [673.415, 739.072] mm. Using Equation (33), the *MSE_h* error is calculated to be 3.80%, with a central distance of approximately 2.44 cm. The obtained result meets the measurement requirements.

As shown in Figure 8, the parameter identification structure indicates that the depth information is [739.05, 811.131] mm. Using Equation (33), the *MSE_h* error is calculated to be 3.88%, with a central distance of approximately 3.09 cm. The obtained result meets the measurement requirements.

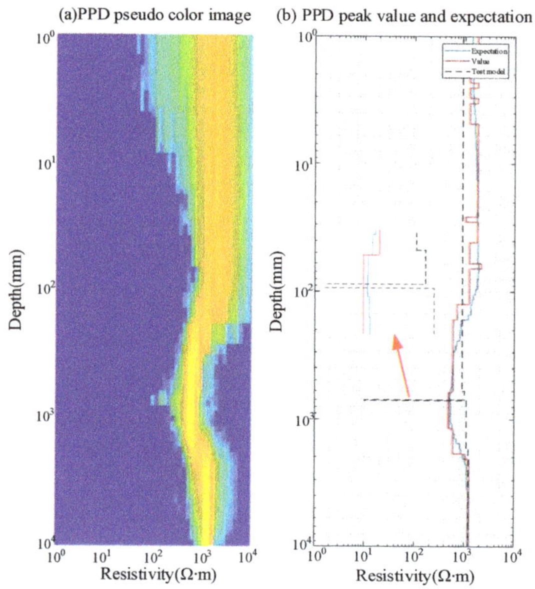

Figure 7. Dimensional Bayesian parameter identification results for grounding grid with depth of 0.7 m.

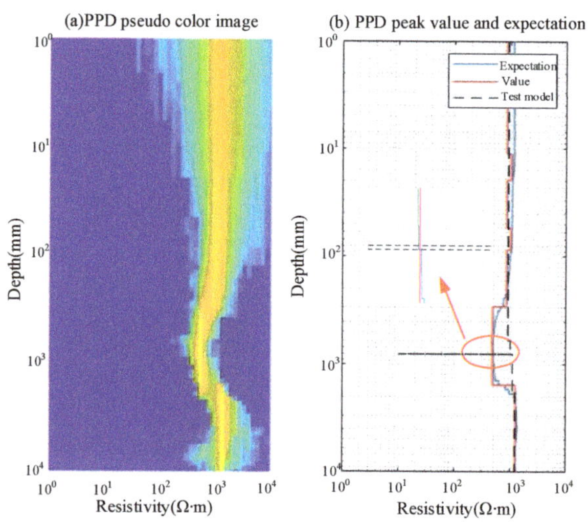

Figure 8. Dimensional Bayesian parameter identification results for grounding grid with depth of 0.8 m.

The parameter identification results for grounding grid depths of 0.7 m and 0.8 m demonstrate that, under slight noise conditions, the error values obtained using Equation (33) are all within 10%, and the midpoint distance between the most likely interval for representing the grounding grid position, identified by the parameter identification, and the midpoint of the interval set in the forward model is also less than 10 cm.

5. Experimental Validation

5.1. Experimental Process

This study presents a signal acquisition study of a substation in a certain region using a pulsed eddy current device. The grounding detection device adopts a concentric integrated design of an emitting coil and a receiving coil. The number of turns of the emitting coil is 50, with a single-turn diameter of 2 mm, and the diameter of the coil frame is 600 mm. The number of turns of the receiving coil is 1400, with a diameter of 0.5 mm. The excitation current of the emitting coil is a 20 A pulse square wave with a frequency of 25 Hz. The sampling rate of the receiving coil is 125 MHz, and the sampling accuracy is 14 bits. The acquisition method is point-based, meaning measurements are taken at intervals of 0.5 m. Each measurement point should capture data for at least 100 cycles. The field measurement setup is shown in Figure 9a. In this experiment, the excitation current is set to 30 A, utilizing a bipolar pulsed current with a frequency of 25 Hz. The coil radius is 0.3 m, and the turn-off time is approximately 10μs. A schematic diagram of the detection device is shown in Figure 9b.

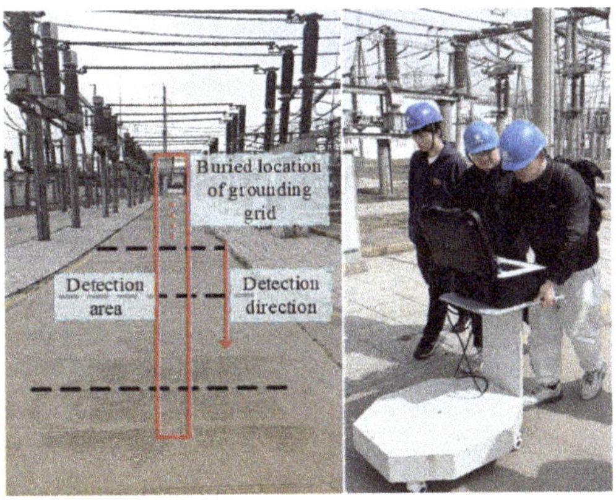

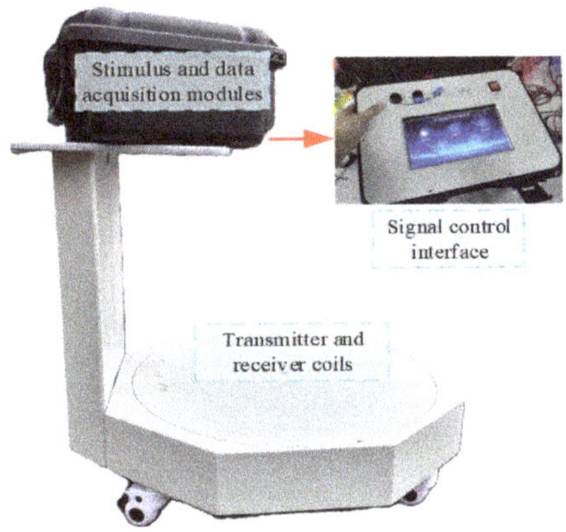

Figure 9. Field signal acquisition and detection device. (**a**) On-site detection schematic diagram. (**b**) Grounding detection device.

The sampling rate of the receiving system is 1 MHz, with a gain factor of 100 times and 300 turns for the receiving coil. According to the Chinese Standard. Electrical Installation Engineering Grounding Device Construction and Acceptance Specification GB 50169-2016, Beijing: China Electricity Council, the soil resistivity and the burial depth of the grounding device should meet the requirements specified in the table below to ensure the grounding resistance design meets the necessary standards.

According to Table 3, it can be observed that when measuring the surface soil resistivity, the variation within a certain depth range is relatively small. Therefore, the measured resistivity can be used as a reference resistivity for the design of the substation grounding system. If the soil resistivity is known, the burial depth range of the grounding grid can be preliminarily estimated and used as prior information for the grounding grid parameter identification calculations. To improve the accuracy of the experiment, prior to its commencement, a four-electrode method was used with a soil resistivity analyzer to measure the soil resistivity near the substation. The measurement results are shown in Figure 10.

Table 3. Soil resistivity and burial depth and maximum length of grounding devices.

ρ ($\Omega \cdot$m)	$100 < \rho \leq 500$	$500 < \rho \leq 1000$	$1000 < \rho \leq 2000$	>2000
Depth (m)	$\geq$0.6	$\geq$0.5	$\geq$0.5	$\geq$0.3
Maximum Length (m)	40	60	80	100

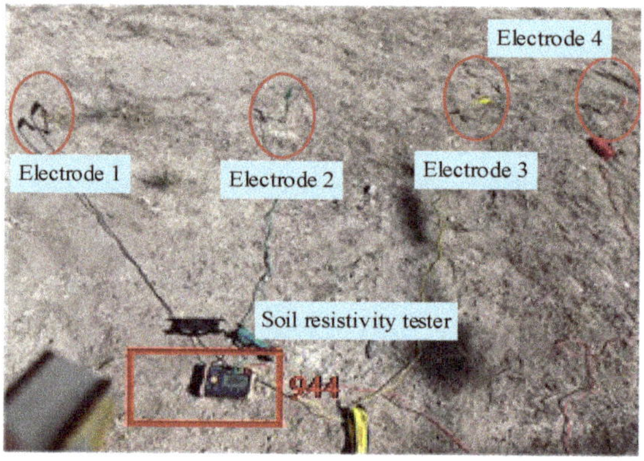

Figure 10. Soil resistivity testing method and results.

During the testing process, the distance between each electrode was set to 1 m. The measured soil resistivity value was 944 $\Omega \cdot$m. According to Table 3, the reasonable burial depth range for the grounding grid should be no less than 0.5 m. This result can serve as prior information for the subsequent grounding grid parameter identification calculations.

5.2. Parameter Identification Results

The variable dimensional Bayesian parameter identification algorithm is employed to calculate the posterior resistivity probability density distribution, which satisfies the data collected and processed through signal processing, as well as the prior information obtained in the previous section. The parameter settings for Bayesian parameter identification are consistent with those in Table 2, with a maximum sampling iteration of one million times and a burn-in sampling of ten thousand times. The model depth range is set between 3 and 10 layers. A total of 10 measurement points are located along the measurement line, with a spacing of 2 m between each point. The multi-point parameter identification results are shown in Figure 11.

In Figure 11, the horizontal axis represents the distance between the 10 measurement points, while the vertical axis indicates the depth information at different points. The color information represents the resistivity of each layer. As shown in the figure, the signal at almost every measurement point shows a variation at the 0.7 m depth layer, indicating the presence of a grounding grid at this depth. The estimated burial depth of the grid is approximately 0.7 m. Excavation verification was then conducted, as shown in Figure 12, where the measured grounding grid burial depth was approximately 0.77 m. The difference between the measured depth and the parameter identification result is 7 cm, with an error of about 9.1%. This demonstrates that the pulsed electromagnetic signal processing and analysis method proposed for substation grounding grid detection in this paper has achieved satisfactory results in detecting the grounding grid depth.

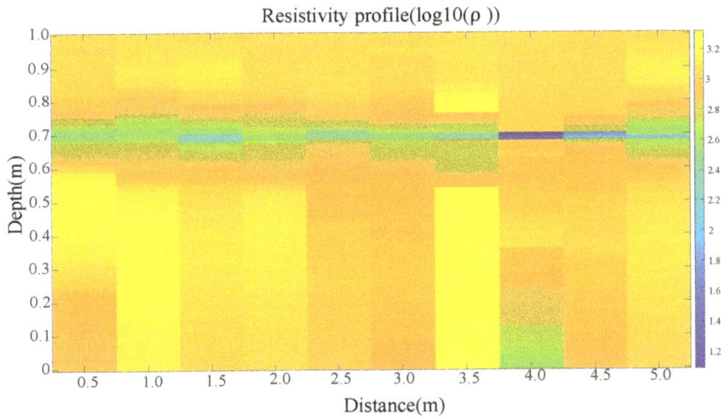

Figure 11. Multi-point Bayesian parameter identification results.

Figure 12. Excavation verification results. (**a**) Field excavation diagram. (**b**) Localized enlarged view.

Through the multiple measurement results of Table 4, it can be seen that the average error between the real results and the parameter identification results is about 7.82%. The parameter identification results proposed in this paper are in good agreement with the actual situation under multiple measurements. The standard deviation of the error of multiple measurement results is 0.0119, which proves the stability of multiple measurement results. In addition, tests were conducted at two additional substations. For soil resistivity measurement, the protocol of the "four-electrode soil resistivity analyzer" described above was followed, with the results presented in Figure 13. The measured soil resistivities of the two substations were 983 Ω·m and 891 Ω·m, respectively; meanwhile, 10 measurement points were arranged in each newly added substation, and consistent pulsed eddy current data acquisition and parameter identification procedures from the original text were adopted to complete the measurement and identification of the grounding grid burial depth. The corresponding results are shown in Tables 5 and 6, respectively.

Table 4. Measurement results of 11 measuring points on site.

Measuring Point	Measured Value z (m)	True value z_{real} (m)	Error Err
1	0.82	0.76	7.89%
2	0.73	0.78	6.41%
3	0.83	0.77	7.79%
4	0.69	0.75	8.00%
5	0.87	0.79	10.13%
6	0.70	0.76	7.89%
7	0.85	0.78	8.97%
8	0.72	0.77	6.49%
9	0.79	0.74	6.76%
10	0.71	0.78	8.97%
11	0.80	0.75	6.67%

(a)

(b)

Figure 13. Substation soil resistivity test results. (**a**) The first substation resistivity measurement result. (**b**) Second substation resistivity measurement result.

Table 5. Measurement results of 10 measuring points on first substation.

Measuring Point	Measured Value z (m)	True Value z_{real} (m)	Error Err
1	1.06	0.98	8.16%
2	0.93	1.01	7.92%
3	1.01	1.11	9.01%
4	1.05	0.96	9.38%
5	0.94	1.02	7.84%
6	1.04	0.96	8.33%
7	1.06	0.97	9.28%
8	0.95	1.03	7.77%
9	0.94	1.04	9.62%
10	0.97	1.06	8.49%

Table 6. Measurement results of 10 measuring points on second substation.

Measuring Point	Measured Value z (m)	True Value z_{real} (m)	Error Err
1	1.29	1.18	9.32%
2	1.31	1.19	10.08%
3	1.11	1.21	8.26%
4	1.12	1.20	6.67%
5	1.32	1.20	10.00%
6	1.27	1.16	9.48%
7	1.09	1.18	7.63%
8	1.12	1.23	8.95%
9	1.11	1.21	8.26%
10	1.10	1.19	7.56%

Derived from the data in Tables 5 and 6, the average values of the identification results for the grounding grid burial depth parameters of the other two substations are 8.58% and 8.69%, respectively. This further demonstrates the reliability and accuracy of the pulsed eddy current parameter identification method based on RJMCMC proposed in this paper under different soil environments.

6. Conclusions

This study addresses the issue of local optima that often arise in parameter identification during substation grounding grid detection using the pulsed eddy current method. A grounding grid burial depth parameter identification method based on variable dimensional Bayesian inference is proposed. First, a numerical forward model of the electromagnetic response of the grounding grid based on magnetic dipoles is derived and subsequently validated. The results demonstrate that the model effectively characterizes the electromagnetic response features of the grounding grid. Second, the RJMCMC method is introduced to enable adaptive adjustment of the model's dimensionality, thereby enhancing the global search capability for parameter identification. A ω-c dual-factor adjustment mechanism is incorporated to improve the resistivity resolution, and a dynamic step size control strategy is adopted to increase computational efficiency. Additionally, a convergence criterion based on uncertainty assessment is introduced, significantly reducing the number of sampling iterations and accelerating the Bayesian parameter identification process. The proposed method is validated using the forward model, and the results show that, under 5% noise interference, the depth information obtained through parameter identification deviates by less than 10 cm from the model depth, with an error of less than 10%. Finally, experimental verification is carried out at a substation in Inner Mongolia, China. The identified grounding grid burial depth is 0.7 m, with a deviation of 9.1% from the actual value, demonstrating the practicality of the proposed parameter identification method.

Author Contributions: Conceptualization, X.K. and Z.L.; methodology, X.K. and J.W.; software, J.H. and S.X.; validation, X.K., Z.L. and J.W.; formal analysis, X.K. and S.X.; investigation, Y.Z. and Z.Z.; resources, Z.Z.; data curation, X.K. and Y.Z.; writing—original draft preparation, X.K.; writing—review and editing, Z.Z.; visualization, J.H. and Z.Z.; supervision, Y.Z.; project administration, S.X.; funding acquisition, J.H. All authors have read and agreed to the published version of the manuscript.

Funding: This research was funded by Inner Mongolia Power (Group) Co., Ltd. Science and Technology Project, Grant No. nmgdlkjxm20240403.

Data Availability Statement: The data used in the analysis presented in the paper will be made available, subject to the approval of the correspondence author.

Conflicts of Interest: The authors declare that this study received funding from Baotou Power Supply Bureau of Inner Mongolia Power Grid Co., Ltd. All authors declare that the research was conducted in the absence of any commercial or financial relationships that could be construed as a potential conflict of interest. The funder was involved in the study design, collection, analysis, interpretation of data, the writing of this article, or the decision to submit it for publication.

References

1. Zhang, B.; He, J.L.; Zeng, R. State of art and prospect of grounding technology in power system. *High Volt. Eng.* **2015**, *41*, 2569–2582. [CrossRef]
2. Xu, X.Y. Research on Fault Diagnosis for Grounding Grid in Substation Based on Tikhonov Regularized Principle. Master's Thesis, Chongqing University, Chongqing, China, 2017.
3. Wang, Y.; Zhang, W.X.; Wang, C.; Li, Z.; Luo, B.; Yan, F.; Zhang, H. Study on numerical simulation of electromagnetic response of intelligent substation grounding grid based on multi-grid partitioning technology. *Electr. Meas. Instrum.* **2024**, *61*, 97–103.

4. Peng, L.S.; Sun, H.Y.; Li, S.S.; Wang, S.; Huang, S.L. Mix-frequency SH guided wave corrosion defects detection method for high voltage transmission tower grounding electrode. *Trans. China Electrotechn. Soc.* **2025**, *40*, 376–386. [CrossRef]
5. Xu, Z.; Liao, X.; Liu, L.; Fu, N.; Fu, Z. Research on Small-Loop Transient Electromagnetic Method Forward and Nonlinear Optimization Inversion Method. *IEEE Trans. Geosci. Remote Sens.* **2023**, *61*, 1–13. [CrossRef]
6. Chen, X.G.; Lei, Y.Z. Analytical expressions of induced coil voltages generated by pulsed eddy currents for metal pipe testing. *Proc. CSEE* **2012**, *32*, 176–182. [CrossRef]
7. Yu, C.; Fu, Z.; Wu, G.; Zhou, L.; Zhu, X.; Bao, M. Configuration Detection of Substation Grounding Grid Using Transient Electromagnetic Method. *IEEE Trans. Ind. Electron.* **2017**, *64*, 6475–6483. [CrossRef]
8. She, S. Thickness measurement and surface-defect detection for metal plate using pulsed eddy current testing and optimized Res2Net network. *IEEE Trans. Instrum. Meas.* **2024**, *73*, 1–13. [CrossRef]
9. Fang, J. Modelling and experimental study on pipeline defect characterisations using a pulsed eddy current measurement. *Nondestruct. Test. Eval.* **2024**, *40*, 4024–4045. [CrossRef]
10. Zeng, Z.W.; Guo, Z.; Gan, F.; Zuo, Y.; Tian, X.; Wang, X.; Lin, Z.; Zhu, W.; Wang, X.; Wang, J. Array coil design and experimental verification for separation of tower grounding pulsed eddy current excitation and response magnetic field signals. *Energies* **2025**, *18*, 364. [CrossRef]
11. Jiang, Y.; Xu, Z.; Wu, P.; Huang, L.; Meng, C. A Time-Domain Calibration Method for Transient EM Field Sensors. *Measurement* **2021**, *168*, 108368. [CrossRef]
12. Wang, H.; Fu, Z.; Wang, Y.; Tai, H.-M.; Chen, W. On-Site Calibration of Air-Coil Sensor for Transient Electromagnetic Exploration. *Geophys. Prospect.* **2019**, *67*, 1595–1610. [CrossRef]
13. Guan, W.J.; Chen, Y.; Wu, W.; Zheng, Q.; Yan, L.; Lu, Y. Fault detection method of substation ground grid based on transient electromagnetic method. In *International Conference of Electrical, Electronic and Networked Energy Systems*; Springer Nature: Singapore, 2024. [CrossRef]
14. Chen, W.Y.; Xue, G.; Lei, K.; Song, W. Transient electromagnetic smoke ring due to a grounded-wire source. *Pure Appl. Geophys.* **2023**, *180*, 2157–2168. [CrossRef]
15. Smiarowski, A.; Hodges, G. Transient electromagnetic smoke rings in a half-space during active transmitter excitation. *Geophysics* **2021**, *86*, E199–E208. [CrossRef]
16. Pasion, L.R. Detecting Unexploded Ordnance with Time Domain Electromagnetic Induction. Ph.D. Thesis, University of British Columbia, Vancouver, BC, Canada, 1999. [CrossRef]
17. Wang, L.; Zhang, S.; Chen, S.; Jiang, H. Fast localization of underground targets by magnetic gradient tensor and Gaussian-Newton algorithm with a portable transient electromagnetic system. *IEEE Access* **2021**, *9*, 148469–148478. [CrossRef]
18. Wang, H.; Chen, S.; Zhang, S.; Yuan, Z.; Zhang, H.; Fang, D.; Zhu, J. A high-performance portable transient electromagnetic sensor for unexploded ordnance detection. *Sensors* **2017**, *17*, 2651. [CrossRef]
19. Chen, S.; Zhang, S.; Zhu, J.; Luan, X. Accurate measurement of characteristic response for unexploded ordnance with transient electromagnetic system. *IEEE Trans. Instrum. Meas.* **2019**, *69*, 1728–1736. [CrossRef]
20. Chen, S.; Zhang, S.; Luan, X.; Liu, Z. Improved differential evolution algorithm for multi-target response inversion detected by a portable transient electromagnetic sensor. *IEEE Access* **2020**, *8*, 208107–208119. [CrossRef]
21. Šimić, M.; Ambruš, D.; Bilas, V. Rapid object depth estimation from position-referenced EMI data using machine learning. *IEEE Sens. J.* **2023**, *23*, 4285–4293. [CrossRef]
22. Šimić, M. Machine Learning Based OBJECT classification from Pulse Induction Metal Detector Data. Ph.D. Dissertation, University of Zagreb, Zagreb, Croatia, 2024.
23. Li, S.; Zhang, X.; Xing, K.; Zheng, Y. Fast inversion of subsurface target electromagnetic induction response with deep learning. *IEEE Geosci. Remote Sens. Lett.* **2022**, *19*, 1–5. [CrossRef]
24. Chen, J.; Yan, F.; Sun, Y.; Zhang, Y. Applicability of transient electromagnetic fast forward modeling algorithm with small loop. *Prog. Electromagn. Res. M* **2020**, *98*, 159–169. [CrossRef]
25. Guptasarma, D. New digital linear filters for Hankel J0 and J1 transforms. *Geophys. Prospect.* **2010**, *45*, 745–762. [CrossRef]

MDPI AG
Grosspeteranlage 5
4052 Basel
Switzerland
Tel.: +41 61 683 77 34

Energies Editorial Office
E-mail: energies@mdpi.com
www.mdpi.com/journal/energies

www.ingramcontent.com/pod-product-compliance
Lightning Source LLC
Chambersburg PA
CBHW061624210926
51850CB00009B/2430
* 9 7 8 3 7 2 5 8 5 3 8 5 4 *